Introduction to Global System for Mobile Communication (GSM)

Physical Channels, Logical Channels, Network, and Operation

Lawrence Harte

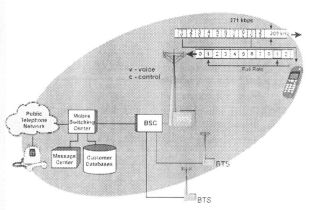

Radio Channel

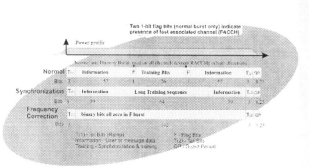

Time Slot Structure

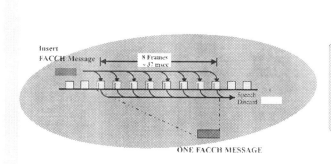

FACCH Signaling

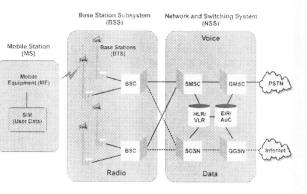

GSM Network

Excerpted From:

Mobile Systems

With Updated Information

ALTHOS Publishing

ALTHOS Publishing

About the Authors

 Mr. Harte is the president of Althos, an expert information provider which researches, trains, and publishes on technology and business industries. He has over 29 years of technology analysis, development, implementation, and business management experience. Mr. Harte has worked for leading companies including Ericsson/General Electric, Audiovox/Toshiba and Westinghouse and has consulted for hundreds of other companies. Mr. Harte continually researches, analyzes, and tests new communication technologies, applications, and services. He has authored over 50 books on telecommunications technologies and business systems covering topics such as mobile telephone systems, data communications, voice over data networks, broadband, prepaid services, billing systems, sales, and Internet marketing. Mr. Harte holds many degrees and certificates including an Executive MBA from Wake Forest University (1995) and a BSET from the University of the State of New York, (1990).

Table of Contents

Introduction to Global System for Mobile Communication (GSM)

Global system for mobile communication (GSM) is a wide area wireless communications system that uses digital radio transmission to provide voice, data, and multimedia communication services. A GSM system coordinates the communication between mobile telephones (mobile stations), base stations (cell sites), and switching systems. Each GSM radio channel is 200 kHz wide channels that are further divided into frames that hold 8 time slots. GSM was originally named Groupe Spéciale Mobile. The GSM system includes mobile telephones (mobile stations), radio towers (base stations), and interconnecting switching systems. The GSM system allows up to 8 to 16 voice users to share each radio channel and there may be several radio channels per radio transmission site (cell site).

Figure 1.1 shows an overview of a GSM radio system. This diagram shows that the GSM system includes mobile communication devices that communicate through base stations (BS) and a mobile switching center (MSC) to connect to other mobile telephones, public telephones, or to the Internet. This diagram shows that the MSC connects to databases of customers. This example shows that the GSM system mobile devices can include mobile telephones or data communication devices such as laptop computers.

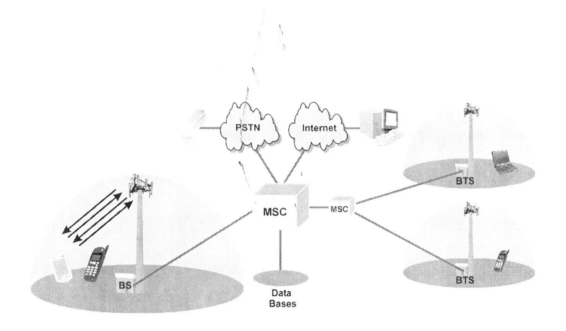

Figure 1.1., Global System for Mobile Communications (GSM)

The GSM specification was initially created to provide a single industry standard for European cellular systems. In 1982, the development of the GSM specification began and the first commercial GSM system began operation in 1991. In 2004, there were more than 1.046 billion GSM subscribers in 205 countries and territories throughout the world [i].

Before the GSM system was available, most countries throughout the world used cellular systems that were often incompatible with each other. Most mobile telephones could only operate on a single type of cellular system so most customers could not roam to neighboring countries. With unique types of systems serving small groups of people, the mass production required to produce low-cost subscriber equipment was not feasible, so subscriber unit equipment costs remained high and early cellular systems enjoyed little success in the marketplace.

At the 1982 Conference of European Posts and Telecommunications (CEPT), the standardization body, Groupe Spéciale Mobile, was formed to begin work on a single European standard. The standard was later named Global System for Mobile Communications (GSM). In 1990, phase one of the GSM specifications were completed, including basic voice and data services. At that time, work began to adapt the GSM specification to provide service in the 1800 MHz frequency range. This 1800 MHz standard, called DCS 1800, is used for the Personal Communications Network (PCN). Phase 2 of the GSM and DCS 1800 specifications, which added advanced short messaging, microcell support services, and enhanced data transfer capability, are now complete. Phase 2+ added advanced information services and packet data transmission capability.

Figure 1.2 shows the basic evolution of the GSM industry standards. This diagram shows that the first release of GSM standards in the early 1990s (phase 1) contained basic voice and data services. The GSM specification was expanded in phase 2 to provide advanced messaging and improved data transfer services. This was followed by phase 2+ of the GSM specification that includes GPRS and EDGE packet data transmission.

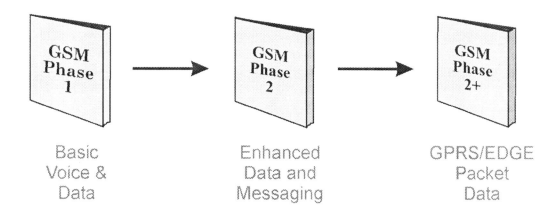

Figure 1.2., Evolution of GSM Standards

The GSM association assists with the promotion, protection, and evolution of GSM technology and products throughout the world. Information about the GSM association can be found at www.GSMWorld.com. GSM association members include mobile operators, manufacturers, and suppliers.

Originally the GSM development group was hosted by (CEPT). GSM technology basics were created in 1987 and in 1989, (ETSI) became the managing body. In 1990, the first GSM specification was released (more than 6,000 pages of specifications). In 1998, the third generation partnership project (3GPP) group was formed to create the next evolution of mobile specification. The 3GPP has now taken over the management of GSM specifications. GSM specifications (and evolved versions of the specification) can be obtained at www.3GPP.org.

GSM Services

The services that GSM can provide include voice services, data services, messaging services, multicast services, and location services.

Voice Services

Voice service is a type of communication service where two or more people can transfer information in the voice frequency band (not necessarily voice signals) through a communication network. Voice service involves the setup of communication sessions between two (or more) users that allows for the real time (or near real time) transfer of voice type signals between users.

The GSM system provides for various types of digital voice services. The voice service quality on the GSM system can vary based on a variety of factors. The GSM system can dynamically change the voice quality because the GSM system can use several different types of speech compression. The service provider can select and control which speech compression process (voice coding) is used. The selection of voice coders that have higher levels of speech compression (higher compression results in less digital bits transmitted) allows the service provider to increase the number of customers it can provide service to with the tradeoff of providing lower quality audio signals. In addition to basic voice services, the GSM system is also capable of providing group voice services and broadcast voice services.

Full Rate Voice

Full rate communication is the dedication of the full capacity of a communication channel to a specific user or application. GSM full rate service allows 8 users to share each radio channel with a voice data rate of 13 kbps for each user.

Half Rate Voice

Half rate communication is a process where only half the normal channel data rate (the full rate) is assigned to a user operating on a radio communications channel. By reducing the data rate, the number of users that can share the radio communications channel can be increased. For the GSM time division multiple access (TDMA) systems, half the number of time slots are assigned during each frame of transmission. This allows other radios to be assigned to the unused time slots. Half rate GSM voice service allows up to 16 users to share each radio channel with a voice data rate of approximately 6.5 kbps for each user.

Enhanced Full Rate Voice

Enhanced full rate (EFR) is an improved form of digital speech compression used in GSM networks. The EFR rate speech coder uses the same data transmission rate as the full rate speech coder. To improve the voice quality, new speech data compression processes (software programs) are used. To use the EFR speech coder, both the mobile station and the system must have EFR capability.

Voice Privacy

Voice privacy is a process of modifying or encrypting a voice signal to prevent the listening of communications by unauthorized users. For digital systems (such as the GSM system,) the digital transmission is modified (encrypted) when a secret key is shared, by both the sender and receiver of the information (voice or data signal). Only users with the secret key can receive and decode the information. The key that is used by the GSM system constantly changes so even if the key is compromised, it cannot be used again.

Voice Group Call Service (Dispatch)

Voice group call service (VGCS) is the process of transmitting a single voice conversation on a channel or group of channels so it can be simultaneously received by a predefined group of service subscribers. VGS allows the simultaneous reception of speech conversation of a predefined group of mobile radios and/or a dispatch console. Each mobile radio that has group call capability is called a group call member.

To help facility communication between multiple mobile devices and to integrate radio communication with other communication systems (such as a computer system,) a dispatch console may be used. A dispatch console is a device or system that allows a person or group of people to access communication systems and services. The person who operates a dispatch console is a dispatcher.

Dispatch consoles can be connected to a group call system wire (such as by an ISDN line) or via a radio base unit. When connected by wire, a dispatch console can be located at any location within or outside a radio coverage area. Specific users (such as a dispatcher) can be assigned priorities to allow them to override the communication of other users.

Group call service is also known as push to talk (PTT) service. PTT is a process of initiating transmission through the use of a push-to-talk button. VGCS operates in half duplex (one-way at a time) communication mode. The push to talk process involves the talker pressing a talk button (usually part of a handheld microphone) that must be pushed before the user can transmit. If the system is available for PTT service (other users in the group not talking), the talker will be alerted (possibly with an acknowledgement tone) and the talker can transmit their voice by holding the talk button. If the system is not available, the user will not be able to transmit/talk.

Each group call member is uniquely identified by their own MSISDN and a group identification number (group ID). Each mobile radio or dispatcher can have access to more than one group code. Calls to a group may be limited to a specific geographic area (specific number of cell sites).

The list of members in a dispatch group along with their identification, assigned priorities, and capabilities is stored in a group call register (GCR).

Figure 1.3 shows how voice group call service may operate in a GSM system. In this diagram, a single voice message is transmitted on GSM radio channels in a pre-defined geographic area. Several mobile radios are operating within the radio coverage limits (group 5 in this example) of the cells broadcasting the group message. In this example, a user is communicating to a group. Each user in this group (including the dispatcher) listens and decodes the message for group 5. Other handsets in the area are not able to receive and decode the group 5 message.

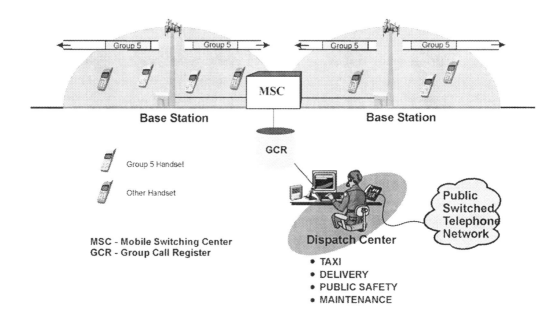

Figure 1.3., GSM Group Call (Dispatch) Service

Voice Broadcast Service (VBS)

Voice broadcast service (VBS) is the process of transferring a single voice conversation or message to be transmitted to a geographic coverage area. VBS subscribers or devices that are capable of identifying and receiving the voice communications then receive the conversation or message.

Figure 1.4 shows the basic operation of voice broadcast service. This example shows how an urgent news message (traffic alert) can be sent to all mobile devices that are operating within the same radio coverage area.

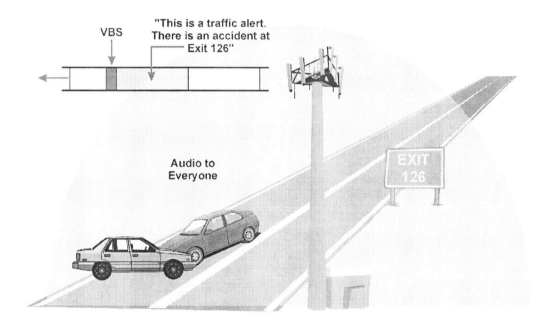

Figure 1.4., Voice Broadcast Service (VBS)

Data Services

Data Services are communication services that transfer information between two or more devices. Data services may be provided in or outside the audio frequency band through a communication network. Data service involves the establishment of physical and logical communication sessions between two (or more) users that allows for the non-real time or near-real time transfer of data (binary) type signals between users.

When data signals are transmitted on a non-digital channel (such as an analog telephone line), a data modem must be used. The data modem converts the data signal (digital bits) into tones that can be transferred in the audio frequency band. Because the speech coder used in the GSM system only compresses voice signals and not data modem signals, analog modem data cannot be sent on a GSM traffic (voice) channel.

When data signals are transmitted on a GSM radio channel, a data transfer adapter (DTA) is used. The DTA converts the data bits from a computing device into a format that is suitable for transmission on a communication channel that has a different data transmission format. DTAs are used to connect communication devices (such as a PDA or laptop) to a mobile device when it is operating on a GSM digital radio channel.

The data services that the GSM system can provide include low-speed circuit switched data to medium speed packet data.

Circuit Switched Data

Circuit switched data is a data communication method that maintains a dedicated communications path between two communication devices regardless of the amount of data that is sent between the devices. This gives to communications equipment the exclusive use of the circuit that connects them, even when the circuit is momentarily idle.

To establish a circuit-switched data connection, the address is sent first and a connection (possibly a virtual non-physical connection) path is established. After this path is setup, data is continually transferred using this path until the path is disconnected by request from the sender or receiver of data.

Packet Switched Data

Packet switched data is the transfer of information between two points through the division of the data into small packets. The packets are routed (switched) through the network and reconnected at the other end to recreate the original data. Each data packet contains the address of its destination. This allows each packet to take a different route through the network to reach its destination.

To provide packet data service, the GSM system uses general packet radio service (GPRS).GPRS is a portion of the GSM specification that allows packet radio service on the GSM system. The GPRS system adds (defines) new packet control channels and gateways to the GSM system.

GPRS packet-switched data service is an "always-on" type of service. When the GSM device is initially turned on, it takes only a few seconds to obtain an IP address that is necessary to communicate with the network. Even when the GSM device is inactive and placed in the dormant state, reconnection is typically less than 1/2 a second.

Fax Services

Fax service is the transmission of facsimile (image) information between users. Facsimile signals have characteristics that are very different than audio signals. As a result, fax transmission involves the use of a communication channel that can send all audio frequencies or a data channel that is setup specifically for the transmission of fax information.

Facsimile signals cannot be sent through the GSM speech coder. This requires the mobile telephone and GSM system to be setup for facsimile transmission. This may be automatically accomplished when a fax machine is connected to a GSM telephone or adapter or it may be manually accomplished through a keypad operation.

Multicast Services

Multicast service is a one-to-many media delivery process that sends a single message or information transmission that contains an address (code) that is designated for several devices (nodes) in a network. Devices must contain the matching code to successfully receive or decode the message. GSM multicast services can include news services or media (digital audio) broadcasts.

Short Messaging Services

Short message service (SMS) gives mobile phone subscribers the ability to send and receive text or data messages. GSM mobile device can send short messages or it can be sent by other systems (such as an email or web page link).

The GSM system limit the short message to 160 alphanumeric characters (7 bits each), 140 data elements (8 bits each), or 70 two type characters (16 bits each). SMS messages can be received while the mobile telephone is in stand-by (idle) or while it is in use (conversation). While the mobile telephone is communicating both voice and message information, short message transfer takes slightly longer than it does while the mobile telephone is in standby. Short messages can be cascaded together to produce longer messages. Short messages are received, stored, and forwarded through the use of a SMS service center (SC).

SMS can be divided into three general categories: Point-to-point, Point-to-multi-point, and broadcast. Point-to-point SMS sends a message to a single receiver. Point to multi-point SMS sends a message to several receivers. Broadcast SMS sends the same message to all receivers in a given area. Broadcast SMS differs from point to multi-point because it places a unique "address" with the message to be received. Only mobile telephones capable of decoding that address receive the message.

Short messages that are received by a mobile telephone are typically stored in the SIM card. This allows the user to keep all their messages on a single SIM card regardless on which mobile telephone they use with the SIM card.

The receipt of short messages can be acknowledged or unacknowledged. Short messages can be setup to request a response such as confirmation of a meeting time or place. Short messages can be originated by the mobile phone, called mobile originated short message service (MOSMS) or by messages may be created by another source, called mobile terminated message service (MTSMS). Because mobile telephones usually have a limited number of keys (compared to a computer), mobile telephones may include predefined messages or use a form of predictive text entry that looks up the possible likely works as portions of the word are completed.

Point to Point Messaging

Point to point messaging is the process of sending data, text or alphanumeric messages from one communication device to one other communication device. To send point to point message, the destination address is selected and added to a message that is sent through the communication network. An example of a point to point message is sending a message to a friend informing them of a place and time to meet.

Point to Multipoint Messaging

Point to multipoint messaging is the process of sending data, text or alphanumeric messages from one communication device to several communication devices. To send point to multipoint messages, a message is copied and sent to each communication device that is listed in the multipoint distribution list. An example of a point to multipoint message is sending a message to a company project team informing them of a change in staff meeting time.

Cell Broadcast Messaging

Cell broadcast messaging is the process of sending SMS messages to all mobile telephones that are operating in the radio coverage of a specific cell site. To send cell broadcast messages, a message is sent to the system operator with instructions to release the message to specific distribution area (one or more cell radio coverage areas). An example of a cell broadcast message is sending a traffic jam message to all people within the area of an automobile accident.

Mobile telephones do not acknowledge receipt of broadcast SMS. If a mobile telephone is performing other tasks (such as scanning for other radio channels) or is turned off, it will miss the broadcast message. To overcome this limitation, the broadcast message may be sent several times. If a mobile telephone has already received the broadcast message, it may ignore the repeated messages.

Executable Messages

An executable message is received by a subscriber identity module (SIM) card in a wireless system (such as a mobile phone system) that contain a program that instructs the SIM card to perform processing instructions.

Flash Messages

Flash SMS automatically displays the SMS message as soon as it is received. An example of a flash message is an important news alert or weather bulletin that is immediately displayed on a mobile telephone display.

Location Based Services (LBS)

Location based services are information or advertising services that vary based on the location of the user. The GSM system permits the use of different types of location information sources including the system itself or through the use of global positioning system (GPS).

GSM Products (Mobile Devices)

GSM mobile devices (also called mobile stations) are voice and/or data input and output devices that are used to communicate with a radio tower (cell sites). GSM end user devices include removable subscriber identity modules (SIMs) that hold service subscription information. The common types of available GSM devices include mobile telephones, PCMCIA cards, embedded radio modules, and external radio modems.

Subscriber Identity Module (SIM)

A subscriber identity module (SIM) is an "information" card that contains service subscription identity and personal information. The SIM card contains at least two numbers that identify the customer; the international mobile subscriber identity (IMSI) and a secret authentication key number K.

The SIM contains a microprocessor, memory and software to hold and process information that includes a phone number, billing identification information and a small amount of user specific data (such as feature preferences and short messages.) This information can be stored in the card rather than programming this information into the phone itself. A SIM card can be either credit card-sized (ISO format) or the size of a postage-stamp (Plug-In format). SIM cards can be inserted into any SIM ready communication device.

Access to a SIM card usually requires the use of a personal identity number (PIN) to restrict access to the SIM card to people who know the code. SIM cards may also be locked to the communication device by a SIM lock code (the service provider only knows the SIM lock code).

The SIM lock code ensures that a communication device will only work with one or a group of subscriber identity module (SIM) cards. The use of a SIM lock code by a service provider helps to ensure that a customer will only be able to use a communication device they provide at low cost with their SIM cards. If another SIM card is inserted to a communication device that is locked to a specific SIM card, the communication device will not operate.

Mobile Telephones

Mobile telephones are radio transceivers (combined transmitter and receives) that convert signals between users (typically people, but not always) and radio signals. Mobile telephones can vary from simple voice units to advanced multimedia personal digital assistants (PDAs). GSM mobile telephones may only include GSM capability (single mode) or it may include GSM and other types of wireless capability (dual mode). GSM mobile device may be only able to receive on one frequency band (single band) or two or more frequency bands (dual band or tri-band).

PCMCIA Air Cards

The PCMCIA card uses a standard physical and electrical interface that is used to connect memory and communication devices to computers, typically laptops. The physical card sizes are similar to the size of a credit card 2.126 inches (51.46 mm) by 3.37 inches (69.2 mm) long. There are 4 different card thickness dimensions: 3.3 (type 1), 5.0 (type 2), 10.5 (type 3), and 16 mm (type 4). GSM PCMCIA radio cards can be added to most laptop computers to avoid the need of integrating or attaching radio devices.

Embedded Radio Modules

Embedded radio modules are self contained electronic assemblies that may be inserted or attached to other electronic devices or systems. Embedded radio modules may be installed in computing devices such as personal digital assistants (PDAs), laptop computers, and other types of computing devices that can benefit from wireless data and/or voice connections.

External Radio Modems

External radio modems are self contained radios with data modems that allow the customer to simply plug the radio device into their USB or Ethernet data port on their desktop or laptop computer. External modems are commonly connected to computers via standard connections such as universal serial bus (USB) or RJ-45 Ethernet connections.

Figure 1.5 shows the common types of GSM products available to customers. This diagram shows that the product types available for GSM include single mode, dual mode and dual frequency mobile telephones, PCMCIA data cards, embedded radio modules, and external radio modems. GSM mobile telephones may be capable of operating on other systems (dual mode) or multiple frequencies. Small radio assemblies (modules) may be inserted (embedded) into other devices such as laptop computers or custom communication devices. PCMCIA data cards may allow for both data and voice operations when inserted into portable communications devices such as laptops or personal digital assistants (PDAs). External modems may be used to provide data services to fixed users (such as desktop computers).

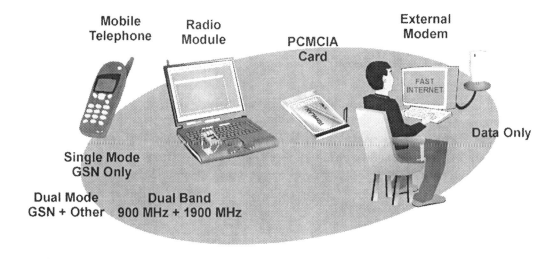

Figure 1.5., GSM Product Types

GSM Radio

GSM radio is wireless communication system that divides geographic areas into small radio areas (cells) that are interconnected with each other. Each cell coverage area has one or several transmitters and receivers that communicate with mobile telephones within its area.

GSM radio systems operate in a specific frequency band (or frequency bands) that has been allocated to the system. The radio frequency channel that the system operators may be reused at different cell sites according to a frequency plan. Users share each radio channel using a combination of frequency division and time division multiple access.

Frequency Allocation

Frequency allocation is the amount of radio spectrum (frequency bands) that is assigned (allocated) by a regulatory agency for use for specific types of radio services.

The original GSM system was assigned two 25 MHz bands at 890-915 MHz (mobile telephone transmit) and 935-960 MHz (base transceiver station transmit) that are separated by 45 MHz. Because each GSM radio channel has a frequency bandwidth of 200 kHz, this divides into 125 radio channel carriers. In some systems, the entire frequency band may not be available, and in other systems, radio channels may be divided among multiple cellular service providers.

Since its creation, many countries have authorized additional frequency band for GSM system. The GSM frequency band for PCN (DCS 1800) is 1710-1785 MHz (subscriber unit transmit) and 1785-1880 MHz (base station transmit) separated by 75 MHz. Each PCN frequency band is divided into 375 radio channels of 200 kHz each.

Figure 1.6 shows the frequency bands that can be used for GSM radio channels. This table shows that GSM systems can operate in the 400 MHz band, 800 MHz band, 900 MHz band, 1800 MHz band, and 1900 MHz frequency bands. This table also shows that each GSM system requires two frequency bands (for duplex operation); one for base to mobile (downlink) and another for mobile to base (uplink). The frequency spacing between downlink and uplink increases as the frequency band increases.

Frequency	Range
GSM400	450.4 - 457.6 MHz paired with 460.4 - 467.6 MHz or 478.8 - 486 MHz paired with 488.8 - 496 MHz
GSM 850	824 - 849 MHz paired with 869 - 894 MHz
GSM900	880 - 915 MHz paired with 925 - 960 MHz
GSM1800	1710 - 1785 MHz paired with 1805 - 1880 MHz
GSM1900	1850 - 1910 MHz paired with 1930 - 1990 MHz

FFigure 1.6., GSM Frequency Bands

Frequency Reuse

Frequency reuse is the process of using the same radio frequencies on radio transmitter sites within a geographic area that are separated by sufficient distance to cause minimal interference with each other. Frequency reuse allows for a dramatic increase in the number of customers that can be served (capacity) within a geographic area on a limited amount of radio spectrum (limited number of radio channels).

Frequency planning is the assignment (coordination) of radio channel frequencies in wireless systems that have multiple transmitters to minimize the amount of interference caused by transmitters that operate on the same frequency. Frequency planning is used to help ensure that combined interference levels from nearby transmitters that are operating on or near the same frequency do not exceed a certain interference (desired signal to interference) level compared to the desired signal.

The ability to reuse frequencies depends on various factors that include the ability of channels to operate in with interference signal energy attenuation between the transmitters. A frequency plan is the assignment of radio frequencies to radio transmission sites (cell sites) that are located within a defined geographic area. The frequency plan may use ratios that are different dependent on the number of transmitting sites to the number of antennas (sectors) on each site. A common frequency reuse plan for GSM is the ability to reuse a radio frequency on every 4th site that has three 120 degree sectors each – 12 total sectors. This plan is commonly called "4/12".

The radio channel bandwidth of GSM carriers are wider than its analog predecessors and the modulation GSM uses is resistant to interfering signals. As a result, GSM radio channels can tolerate interfering signals up to 20% (9 dB below) of the desired signal compared to analog signals that can only tolerate 1.6% to 6.3 % of (18-12 dB below) their received signal [ii].

Figure 1.7 shows how GSM can use frequency reuse to increase the system capacity. This diagram shows that a frequency in a GSM system can be reused at nearby cell sites provided the radio signal level from the interfering (unwanted) cell is 9 dB to 14 dB below the desired signal level.

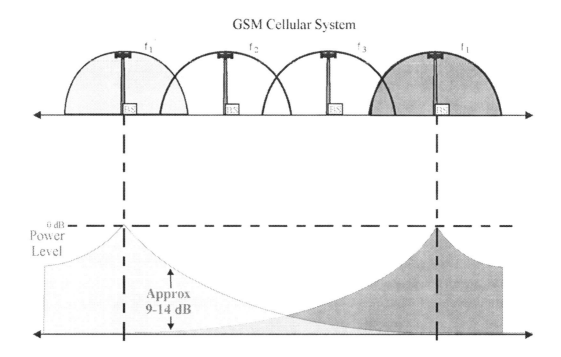

Figure 1.7., GSM Frequency Reuse

Time Division Multiple Access (TDMA)

Time division multiple access (TDMA) is a process of sharing a single radio channel by dividing the channel into time slots that are shared between simultaneous users of the radio channel. When a mobile radio communicates with a TDMA system, it is assigned a specific time position on the radio channel. By allow several users to use different time positions (time slots) on a single radio channel, TDMA systems increase their ability to serve multiple users with a limited number of radio channels.

GSM uses time division multiplexing (TDM) to share one modulated carrier frequency radio waveform among 8 (full rate) to 16 (half rate) conversations. Therefore, documents related to GSM are careful to distinguish between a radio carrier and a communication channel.

Figure 1.8 shows how the GSM system allows more than one simultaneous user per radio channel through the use of time multiplexing. This example shows GSM radio channel can be divided to allow 8 or 16 users per channel. The top example shows that one slot per frame is assigned to full rate users. The bottom example shows that one slot for every other frame is assigned to half rate users.

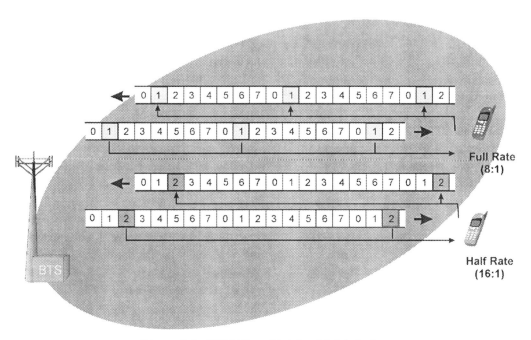

Figure 1.8., GSM Time Division Multiplexing

RF Power Control

RF power control is a process of adjusting the power level of a mobile radio as it moves closer and further away from a transmitter. RF power control is typically accomplished by the sensing of the received signal strength level and the relaying of power control messages from a transmitter to the mobile device with commands that are used to increase or decrease the mobile device's output power level. GSM RF power adjustments occur in 2 dB steps.

The use of RF power control allows for the transmission of only the necessary RF signal level to maintain a quality communication link. Some of the key benefits of RF power control is reduced radio channel interference to other radio devices and increased batter life.

Figure 1.9 shows how the radio signal power level output of a mobile telephone is adjusted by commands received from the base station to reduce the average transmitted power from the mobile telephone. This lower power reduces interference to nearby cell sites. As the mobile telephone moves closer to the cell site, less power is required from the mobile telephone and it is commanded to reduce its transmitter output power level. The base station transmitter power level can also be reduced although the base station RF output power is not typically reduced. While the maximum output power varies for different classes of mobile telephones, typically they have the same minimum power level.

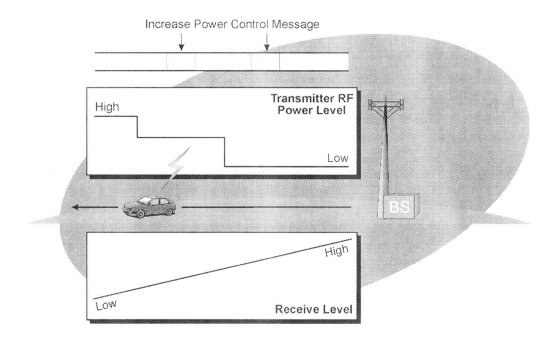

Figure 1.9., RF Power Control

Mobile radios can be classified by the maximum RF power they can transmit. RF Power classification defines the RF power levels associated with specific modes of operation for a particular class of radio device. Classes of RF devices often vary based on the application and use of the device such as portable, mobile or fixed applications. RF power classification typically defines the maximum RF power level a device may transmit but it may also include the minimum RF power levels and the RF power levels for specific modes of operation (such as during a radio transmission burst).

There are 5 different RF power classes used for mobile telephones in the GSM system, class 1 through class 5. Class 1 devices can transmit up to 20 Watts (+43 dBm), class 2 can transmit up to 8 Watts, class 3 can transmit up to 5 Watts, class 4 can transmit up to 2 Watts, and class 5 can transmit up to 0.8 Watts.

During normal operation, the mobile device uses 1 slot out of 8 so the average power is $1/8^{th}$ of the transmitted power. This means a class 4 device that is transmitting at its maximum power of 2 Watts is actually only transmitting 250 mWatts ($1/8^{th}$ of 2 Watts). Base stations continuously transmit regardless if all the time slots are used so their average transmitter RF power is the same as their peak transmit power.

Figure 1.10 shows the different types of power classes available for GSM mobile devices and how their maximum power level. This table shows that there are 5 classes of GSM mobile devices and their maximum power level ranges from 0.8 Watts to 20.0 Watts.

Power Class	Mobile Station Power (Watts)
1	20
2	8
3	5
4	2
5	0.8

Figure 1.10, GSM RF Power Classification

Mobile Assisted Handover (MAHO)

Mobile assisted handover is a process that is used to allow a mobile phone to assist in the base station in the decision to transfer the call (handoff/handover) to another base station. The mobile radio assists by providing RF signal quality information that typically includes received signal strength indication (RSSI) and bit error rate (BER) of its own and other candidate channels. MAHO is an official term of the GSM system.

During GSM communication, the mobile transmits on one slot, receives on one slot, and has 6 idle slots available in each frame. During the idle time periods, the mobile telephone can tune to other radio channel frequencies and measure their signal strength.

Figure 1.11 illustrates the basic mobile assisted handover process. The mobile telephone initially receives a list of nearby radio channels to monitor. During the idle of the mobile telephone periods (between transmission and reception bursts), the mobile telephone monitors other radio channels for signal strength. The mobile telephone can report these measurements along with its own received signal strength and channel quality (bit error rate) back to the base station. The base station can use this information along with other information to determine if a new radio channel should be assigned and which channel to assign the mobile telephone to.

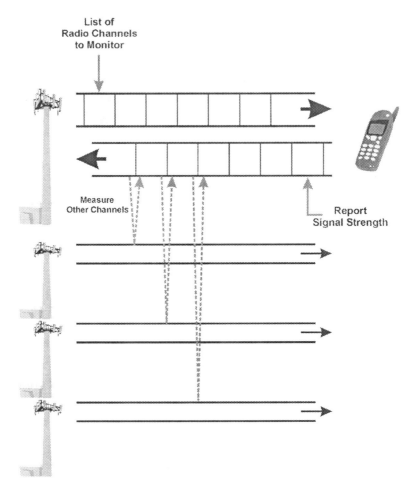

Figure 1.11., Mobile Assisted Hand-over

Digital Audio and Baseband

Digital audio is the representation of audio information in digital (discrete level) formats. The use of digital audio allows for more simple storage, processing, and transmission of audio signals. Baseband audio processing includes analog to digital conversion, digital Speech compression, and channel coding.

Analog to Digital Conversion (ADC)

Analog to digital conversion is a process (digitization) that changes a continuously varying signal (analog) into digital values. The GSM system converts analog audio signals into digital form so it can be compressed and coded onto the radio channel.

A typical analog to digital conversion process includes an initial filtering process to remove extremely high and low frequencies that could confuse the digital converter. This is followed by a periodic sampling section that measures the instantaneous level of the signals at fixed time intervals and converts the measured values (sampled voltages) into its equivalent digital number or pulses.

Figure 1.12 shows how an analog signal is converted to a digital signal. This diagram shows that an acoustic (sound) signal is converted to an audio electrical signal (continuously varying signal) by a microphone. This signal is sent through an audio band-pass filter that only allows frequency ranges within the desired audio band (removes unwanted noise and other non-audio frequency components). The audio signal is then sampled every 125 microseconds (8,000 times per second) and converted into 8 digital bits. The digital bits represent the amplitude of the input analog signal.

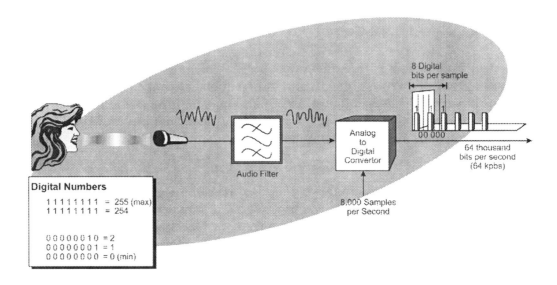

Figure 1.12., Analog to Digital Conversion

Digital Speech Compression (Speech Coding)

Digital speech compression (speech coding) is a process of analyzing and compressing a digitized audio signal, transmitting that compressed digital signal to another point, and decoding the compressed signal to recreate the original (or approximate of the original) signal.

The GSM digital speech compression process works by grouping the 64 kbps digital audio signals into 20 msec speech frames. These speech frames are analyzed and characterized (e.g. volume, pitch) by the speech coder. The speech coder removes redundancy in the digital signal (such as silence periods) and characterizes digital patterns that can be made by the human voice using code book tables. The code book table codes are transmitted instead of the original digitized audio signal. This results in the transmission of a 13 kbps compressed digital audio instead of the 64 kbps digitized audio signal.

Figure 1.13 shows the basic speech data compression process used for the GSM speech coder. This diagram shows that the analog voice signal is sampled 8,000 times each second and digitized into a 64 kbps digital signal. The digitized signal is grouped into 20 msec speech frames. The speech frames are analyzed and compressed into a new 13 kbps digital signal.

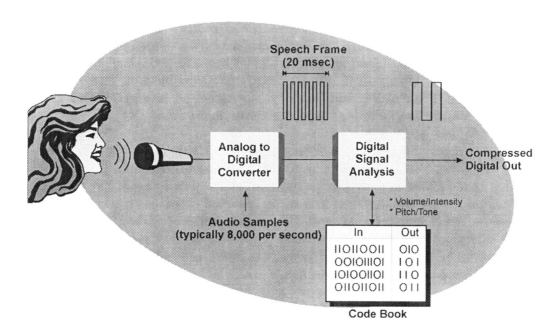

Figure 1.13., Digital Speech Compression

There are several types of speech coding that can be used in GSM systems and devices. The first generation of speech coding was Regular Pulse Excitation-Long Term Prediction (RPE-LTP). Since the first GSM speech coder was developed in 1988, speech coding technology has improved and this had lead to the introduction of a new enhanced full rate (EFR) speech coder. The EFR provides improved voice quality using the same 13 kbps data transmission rate. If the mobile telephone and the system both have the EFR speech coder available, it can be used.

Channel Coding

Channel coding is a process where one or more control and user data signals are combined with error protected or error correction information. After a sequence of digital data bits has been produced by a digital speech code or by other digital signal sources, these digital bits are processed to create a sequence of new bit patterns that are ready for transmission. This processing typically includes the addition of error detection and error protection bits along with rearranging of bit order for transmission.

The error protection and control bits increase 13 kbps user data transmission rate to 22.8 kbps. In addition to adding error protection bits, the data that is transmitted is distributed (interleaved) over 8 adjacent slot periods. This allows only some of the bits to be received in error if a transmitted packet is lost (due to burst errors). Using the error protection coding, it may be possible to recreate (replace) these bits. The GSM system uses several types of error protection coding including cyclic redundancy check (CRC), block code, and convolutional coding.

Cyclic Redundancy Check Sum (CRC)

Cyclic redundancy check is an error-checking process in which bytes at the end of a packet are used by the receiving node to detect transmission problems. The bytes represent the result of a calculation performed on the data portion of the packet before transmission. If the results for the same calculation on the received packet are not equal to the transmitted results, the receiving node can request that the packet be re-sent.

In the GSM system, CRC error protection codes are used in all call processing messages. CRC error protection codes are also used for some of the more important speech coding bits (not all of them).

Block Code

Block codes are a series of bits or a number that is appended to a group of bits or batch of information that allows for the detecting and/or correcting of information that has been transmitted. Block codes use mathematical formulas that perform an operation on the data that will be transmitted. This produces a resulting number that is related to the transmitted data. Depending on how complex the mathematical formula is and how many bits the result may be, the bock code can be used to detect and correct one or more bits of information.

Convoultional Coding

Convolutional coding is an error correction process that uses the input data to create a continuous flow of error protected bits. As these bits are input to the convolutional coder, an increased number of bits are produced. Convolutional coding is often used in transmission systems that often experience burst errors such as wireless systems. Convolutional coding systems are represented by the ratio (rate) of input bits to output bits. A ½ rate convolutional coder creates (outputs) 2 bits for each 1 bit (input) it receives.

Echo Cancellation

Echo cancellation is a process of extracting an original transmitted signal from the received signal that contains one or more delayed signals (copies of the original signal). Echoes may occur as a result of transmission delays in the audio signal and through acoustic feedback where some of the audio signal transferring from a speaker into a microphone. Echoed signals cause distortion and may be removed by performing via advanced signal analysis and filtering. The specific process of echo canceling that is used (if any) is not specified.

Figure 1.14 shows how echoes can be removed. In this example, the transmission of the words: "Hello, is Susan there" experience the effects of echo. When the signal is supplied to an echo canceller (a sophisticated estimating and subtraction machine), the echo canceling device takes a sample of the initial audio and tries to find echo matches of the input audio at delayed

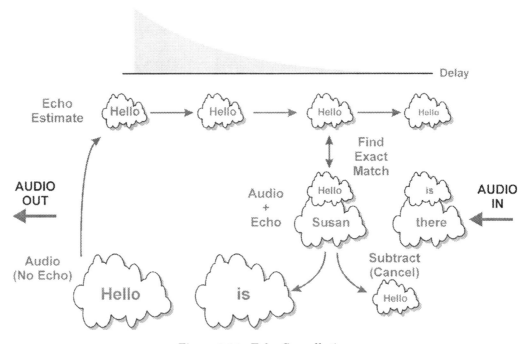

Figure 1.14., Echo Cancellation

periods (the amount of echo time). In this example, it does this by creating various delayed versions the audio signal and different (reduced) amplitude (echo volume usually decreases as time increases), and comparing the estimate the audio that contains the echo. When it finds an exact match at a specific audio level, the echo canceller can subtract the echo signal. This produces audio without the echo.

Radio Channels

A radio channel is a communications channel that uses radio waves to transfer information from a source to a destination. A radio channel may transport one or many communication channels and communication circuits.

Channel Bandwidth

Radio channel bandwidth is the difference between the upper frequency limit and lower frequency limit of allowable radio transmission energy for a radio communication channel. The GSM radio channel has a 200 kHz channel bandwidth.

Modulation

Modulation is the process of changing the amplitude, frequency, or phase of a radio frequency carrier signal (a carrier) to change with the information signal (such as voice or data). Digital modulation is the process of creating an analog signal that represents digital information.

The GSM physical radio channels use Gaussian minimum shift keying (GMSK). GMSK is a form of two-level digital FM modulation. The radio channel has a gross data transmission rate of a GSM channel is 271 kbps.

Duplex Channels

Duplex communication is the transmission of voice and/or data signals that allow simultaneous 2-way communication. To provide duplex communication on analog systems, each voice path was assigned to a different transmitter and frequency. This process of using two frequencies for duplex communication is called frequency division duplex (FDD). Another method that can be used for duplex communication is time division duplex (TDD). TDD provides two way communications between two devices by time sharing. When using TDD, one device transmits (device 1), the other device listens (device 2) for a short period of time. After the transmission is complete, the devices reverse their role so device 1 becomes a receiver and device 2 becomes a transmitter. The process continually repeats itself so data appears to flow in both directions simultaneously.

The GSM system uses a combination of FDD and TDD communication. One frequency is used to communicate in one direction and the other frequency is required to communicate in the opposite direction. However, the GSM system also uses TDD as the transmitter and receiver communicate at different times. The time offset of transmission and reception simplifies the design of the mobile device (less radio filter parts).

The radio frequency separation between the forward (downlink) and reverse (uplink) frequencies varies based on the frequency band. In general, the higher the frequency, the larger the frequency separation between the forward and reverse channels. For GSM 900 MHz, the frequency separation is 45 MHz, for PCN the frequency separation is 95 MHz and for GSM PCS 1900 MHz the frequency separation is 80 MHz.

Figure 1.15 shows the frequency and time offsets between the forward and reverse channel for the GSM system. This diagram shows that the frequency offset varies with the system it is operating on. For GSM 900, the frequency separation is 45 MHz, for PCN 1800 the frequency separation is 95 MHz and for the PCS 1900 system, the frequency separation is 80 MHz. This example also shows that the downlink channel is time offset from the uplink channel. This time offset allows the mobile device to transmit at a different time than it receives.

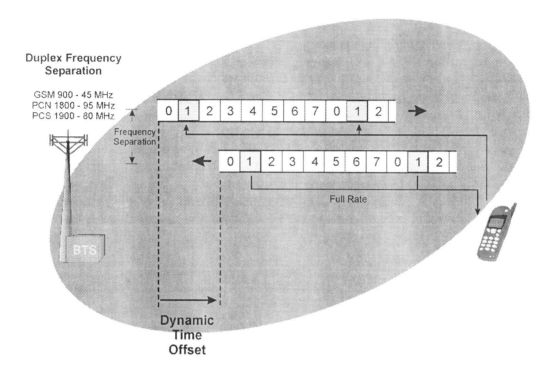

Figure 1.15., GSM Duplex Radio Channels

Radio Channel Structure

Radio channel structure is the division and coordination of a radio communication channel (wireless information transfer) into logical channels, frames (groups) of data, and fields within the frames that hold specific types of information.

The radio channel is divided into frames with 8 time slots per frame (0 through 7) and time slots are divided into field dependent on the purpose of the time slot. A forward (downlink) radio channel is paired with a reverse (uplink) radio channel to provide simultaneous two-way (duplex) voice communication.

Several logical channels can exist on time slots in the physical radio channels. When a radio channel has a control channel, time slot 0 of the frame is used. The other time slots are used for user data. For normal (full rate) voice communication, a time slot in each frame is dedicated for the entire duration of the call. For efficient (half rate) voice communication, a time slot in every other frame is dedicated for the duration of the call. For packet data communication (using GPRS), the time slots are dynamically assigned.

Figure 1.16 shows that the GSM system uses a single type of radio channel. Each radio channel in the GSM system has a frequency bandwidth of 200 kHz and a data transmission rate of approximately 271 kbps. This example shows that each radio communication channel is divided into 8 time slots (0 through 7). This diagram shows that a simultaneous two-way voice commu-

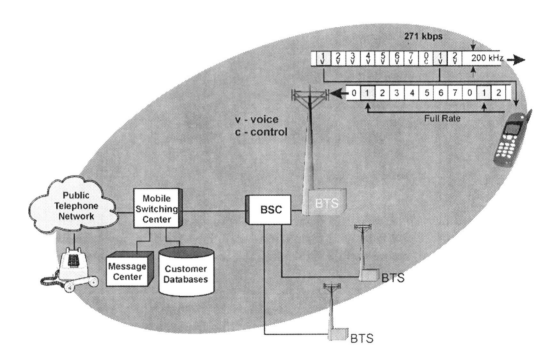

Figure 1.16., GSM Radio Channels

nication session requires at least one radio channel communicates from the base station to the mobile station (called the forward channel) and one channel communicates from the mobile station to the base station (called the reverse channel). This example also shows that some of the radio channel capacity is used to transfer voice (traffic) information and some of the radio channel capacity is used to transfer control messages.

Time Slot Structure

Time slots are the smallest division of a communication channel that is assigned to particular users in a communication system. Time slots can be combined for a single user to increase the total data transfer rate available to that user. In some systems, time slots are assigned dynamically on an as-needed basis.

Slot structure is the division of a time slot into different fields (information) parts. Slot structure fields typically include a preamble for synchronization, control header (e.g. address information), user data, and error detection.

The time period for a GSM time slot is 577 microseconds. The number of data bits in a time slot depending on the type of the time slot (user data or control). The structure of the time slot can also vary dependent if the time slot is on the uplink or downlink radio channel. Each normal time slot contains 148 bits of information. Some time slot data bits are used for user data and other bits are dedicated for control.

The time slots are numbered from 0 to 7. For voice communication, users have a fixed assignment of particular time slots. For packet data transmission (such as GPRS), time slots are dynamically assigned.

Time slots include ramp up and ramp down periods to minimize rapid changes in radio transmitter power. The ramp up and ramp down time is used to reduce unwanted radio emissions that occur from rapidly changing signals.

A single time slot transmission is called a radio burst. Four types of radio bursts are defined in the GSM system. Normal burst, shortened burst, frequency correction burst, and synchronization burst.

Normal Burst

A normal burst is a 577 usec transmission period that is used for normal communication between the mobile device and the base station. Each normal burst can transfer 114 bits of user information data (after error protection is removed).

The first 3 bits of the normal burst time slot are used for the ramp period that allows for the gradual increase in transmitter power level and to send tail bits that are used as part of the convolutional (continuous) error protection channel coding process. Convolutional error coding requires several bits to start the error protection coding process.

A portion of the data bits (57 bits) follow the tail bits. This is followed by a stealing flag that indicates if the normal burst contains user data or if it contains a control message (FACCH message). A sequence of pre-defined training bits (26) are located in the center of the normal burst to assist in the reception and decoding of the bits of the normal burst. The same training bit pattern is used in all eight time slots. This allows mobile telephones to distinguish between their radio channel and other radio channels that are operating on the same frequency from nearby cells. If the mobile device decodes the training bit pattern and it does not match what it is expecting, it should discard the packet. The last 3 bits of the burst are dedicated to the ramp down period.

At the end of the time slot, time is allocated as a guard period when no transmission occurs. The guard period is included to help ensure that transmitted bursts from one mobile device do not overlap transmission bursts from other mobile devices.

Introduction to GSM

Figure 1.17 shows the different types of transmit bursts used in the GSM system and their structure. This example shows that the GSM system includes several burst types; normal burst, synchronization burst, frequency correction burst, and a shortened (access burst). The standard slot time period for a transmit burst is 577 usec long and it contains 156.25 bit periods. The information fields included in the normal bursts include initial tail bits (TB), data bits (D), stealing flags (S), a training sequence (T), and final tail bits (TB). A guard period (GP) is included at the end of the normal burst time period to help ensure that transmitted bursts from one mobile device does not overlap with transmitted bursts from another mobile device. The synchronization burst includes a long training sequence in addition to the synchronization information. The frequency correction burst contains all 0 bits. The shortened access burst contains a tail bits, synchronization bits, and an encrypted access code.

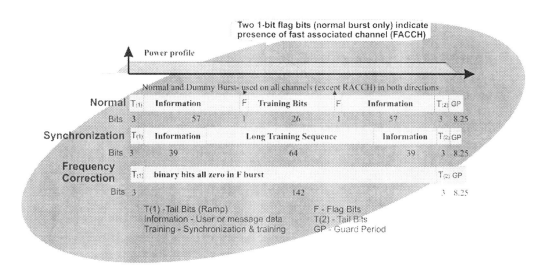

Figure 1.17., GSM Burst Slot Structures

Random Access Burst (Shortened Burst)

A random access burst (shortened burst) is a short 88 bit transmission burst that is used to request access to the GSM system. Mobile devices used a shortened burst when transmitting an access request to the GSM system to avoid the possibility of burst overlap with transmission bursts in adjacent time slots. Once the GSM system has acknowledged the request for service and provides a relative timing adjustment, it can adjust its transmission timing (relative to the received time slots) and begin to transmit normal (full size) time slots.

Mobile telephones may also transmit a shortened burst during handover when the distance between the mobile device and the base station is not known. It is possible to perform handover (even in large cells) without transmitting shortened bursts by allowing the mobile device to synchronize with the cell of the new cell site. It can accomplish this by monitoring the control channel of the new cell site during its idle time periods and acquiring the channel timing information (synchronization information).

Frequency Correction Burst

A frequency correction burst is a time slot of information that contains a 142 bit pattern of all 0 values. The reception and decoding of the frequency correction burst allows the mobile device to adjust (frequency correct) its timing so it can better receive and demodulate the radio channel.

Synchronization Burst

A synchronization burst is a transmission burst that contains system timing information. It contains a 78 bit code to identify the hyperframe counter. The synchronization burst follows the frequency correction burst.

Frame Structure

Frame structure is the division of defined length of digital information into different fields (information) parts. Frame structure fields typically include a preamble for synchronization, control header (e.g. address information), user data, and error detection. A frame may be divided into multiple time slots. The GSM system

A GSM frame is 4.615 msec and it is composed of 8 time slots (numbered 0 through 7). During voice communication, one user is typically assigned to each time slot within a frame.

Between the downlink channel and uplink channel, the time slot numbers are offset by 3 slots. This allows the mobile telephone to transmit at different times than it receives. This allows the design of the mobile device to be simplified by replacing a frequency filter (duplexer) with a more efficient transmit/receive (T/R) switch.

MultiFrame Structure

Multiframes are frames that are grouped or linked together to perform specific functions. Multiframes on the GSM system use established schedules for specific purposes such as coordinating frequency hopping patterns. Multiframes used in the GSM system include the 26 traffic multiframe, 51 control multiframe, superframe, and hyperframe.

Traffic Multiframe Structures

The 26 traffic multiframe structure is used to send information on the traffic channel. The 26 traffic multiframe structure is used to combine user data (traffic), slow control signaling (SACCH), and an idle time period. The idle time period allows a mobile device to perform other necessary operations such as monitoring the radio signal strength level of a beacon channel from other cells. The time interval of a 26 frame traffic multiframe is 6 blocks of speech coder data (120 msec).

Control Multiframe Structures

The 51 control multiframe structure is used to send information on the control channel. The 51 frame control multiframe is sub divided into logical channels that include the frequency correction burst, the synchronization burst, the broadcast channel (BCCH), the paging and access grant channel (PACCH), and the stand-alone dedicated control channel (SDCCH).

Superframe

A superframe is a multiframe sequence that combines the period of a 51 multiframe with 26 multiframes (6.12 seconds). The use of the superframe time period allows all mobile devices to scan all the different time frame types at least once.

Hyperframe

A hyperframe is a multiframe sequence that is composed of 2048 superframes and is the largest time interval in the GSM system (3 hours, 28 minutes, 53 seconds). Every time slot during a hyperframe has a sequential number (represented by an 11 bit counter) that is composed of a frame number and a time slot number. This counter allows the hyperframe to synchronize frequency hopping sequence, encryption processes for voice privacy of subscribers' conversations.

Figure 1.18 shows the different types of GSM frame and Multiframe structures. This diagram shows that a single GSM frame is composed of 8 time slots. When a radio channel is used to provide a control channel, time slot 0 and the other time slots are used for traffic channels. Fifty one frames are grouped together to form control multiframes (for the control channel). Twenty six frames are grouped together to form traffic Multiframes (for the traffic channels). Superframes are the composition of 26 control multiframes or 51 traffic Multiframes to provide a common time period of 6.12 seconds. Two thousand forty eight Superframes are grouped together to form a Hyperframe. A Hyperframe has the longest time period in the GSM system of 3 hours, 28 minutes, and 53 seconds.

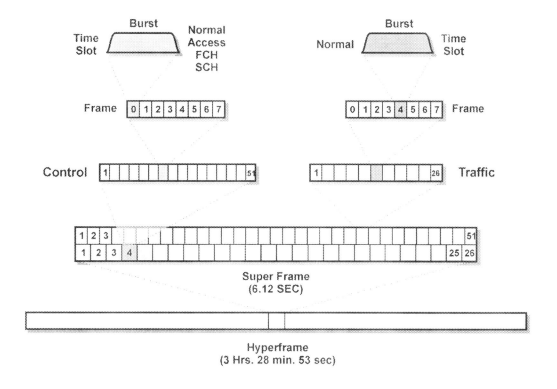

Figure 1.18., GSM Basic Frame and Multiframe Structure

Figure 1.19 shows the GSM time intervals. This table shows that key time periods in the GSM system range from 3.69 usec for a single bit period to 3 hours, 28 minutes, and 53 seconds for a hyperframe period.

	Interval	Notes
Hyperframe	3 hours, 28 minutes, 53 seconds	2048 Superframes
Superframe	6.12 seconds	51 traffic multiframes or 26 control multiframes
Traffic Multiframe	120 msec	Six 20 msec speech frames
Control Multiframe	235.4 msec	51 frames
Frame	4.615 msec	8 time slots
Time Slot	577 usec	217 pulse repetition rate
Bit	3.69 usec	271 kbps

Figure 1.19., GSM Time Intervals

Slow Frequency Hopping

Slow frequency hopping is a process of changing the radio frequencies of a communications on a regular basis (pattern). The duration of transmission on a single frequency is typically much longer than the amount of time it takes to send several bits of digital information. Slow frequency hopping is used to reduce the effects of radio signal fading and to minimize the effects of interference from radio channels that are operating on the same frequency.

Radio signal fading is often limited to a specific frequency range. Radio frequencies that are separated by more than 1 MHz may not fade simultaneously [iii]. If successive time slot bursts are transmitted on different frequencies, if a radio signal fade occurs, it will not likely occur on consecutive bursts.

The effects of radio signal interference that are received from nearby cell sites that operate on the same frequency can be reduced by using slow frequency hopping. Interfering radio signals may only affect particular time slots. Because frequency hopping is combined with error protection that is distributed over multiple time slots (which the GSM system does), a signal fade will produce a lower number of bit errors

The hopping sequence pattern is created by the radio system by assigning a hopping sequence number (HSN) and a mobile allocation index offset (MAIO). The combination of these variables selects a hopping pattern and where the mobile device should be operating within the hopping pattern.

Figure 1.20 shows a simplified diagram of how a slow frequency hopping system transfers information (data) from a transmitter to a receiver using many communication channels. This diagram shows a transmitter that has a preprogrammed frequency tuning sequence and this frequency sequence occurs by hopping from channel frequency to channel frequency. To receive information from the transmitter, the receiver uses the exact same hopping sequence. When the transmitter and receiver frequency hopping sequences occur exactly at the same time, information can transfer from the transmitter to the receiver. This diagram shows that after the transmitter hops to a new frequency, it transmits a burst of information (packet of data). Because the receiver hops to the same frequency, it can receive the packet of data each time.

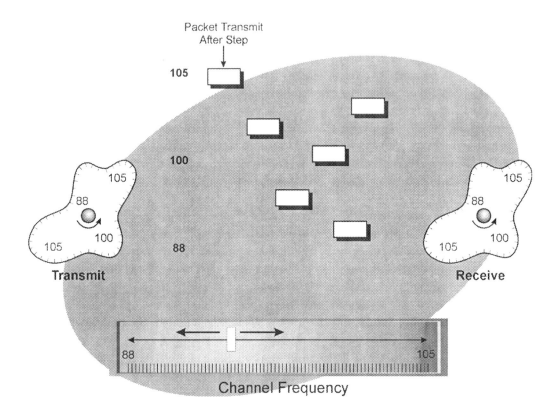

Figure 1.20., Slow Frequency Hopping Example

Discontinuous Reception (Sleep Mode)

Discontinuous reception (DRx) is a process of turning off a radio receiver when it does not expect to receive incoming messages. For DRx to operate, the system must coordinate with the mobile radio for the grouping of messages. The mobile device will wake up during scheduled periods to look for its messages. This reduces the power consumption that extends battery life. This is sometimes called sleep mode.

The WCDMA system divides the paging channel into sub-channel groups to provide for DRx capability. The number of sub-channel groups is determined by the system. Each 10 frames contain a paging channel frame. To inform the mobile device of the sleep periods, a paging indicator channel (PICH) is used. A paging indicator (PI) message is sent at the beginning of the paging channel frame to identify the paging channel group. This allows the mobile device to quickly determine if it must keep its receiver on during the paging group or if it can turn off its receiver and wait for the next paging channel group.

The number of the paging sub-channel is determined by the last digits of the mobile telephone's international mobile service identity (IMSI). The system parameter information sent on the BCCH identifies the grouping of paging sub-channels. The broadcast control channel (BCCH) identifies which multiframes contain paging and access messages and which contain sub-paging classes.

Mobile telephones only need to wakeup for multiframes that are part of its paging sub-channel. During multiframes that are not part of its paging sub-channel, the mobile telephone can set an electronic timer to allow receiver and transmitter circuits to be turned off until the next multiframe group that may contain paging or control messages. The GSM sleep period ranges from approximately 1 to 20 seconds.

Discontinuous Transmission (DTx) Operation

Discontinuous transmission is the ability of a mobile device or communications system to inhibit transmission when no activity or reduced activity is present on a communications channel. DTx is often used in mobile telephone systems to conserve battery life of portable mobile telephones.

The GSM system allows the mobile device to use DTx by intermittently stopping transmission during periods of low audio speech activity. Speech activity is determined by voice activity detection (VAD). When the VAD determines that there is no speech activity, it can temporarily shut off the speech coder and inhibit the transmitter. To ensure the listener does not feel uncomfortable with complete silence periods, a background noise signal may be sent. This "comfort noise" is sent to minimize the change in background noise during inactive voice. During the silence period, the mobile device may continue to compress the background noise and create sent silence descriptor (SID) frames that are sent at a data rate of 500 bps. This small amount of data approximates the same background noise during the silence periods as occurs during normal speech periods. This provides for more uniform communication between the users.

Dynamic Time Alignment

Dynamic time alignment is a technique that allows a radio system base station to receive transmitted signals from mobile radios in an exact time slot, even though not all mobile telephones are the same distance from the base station. Time alignment keeps different mobile radio transmit bursts from colliding or overlapping. Dynamic time alignment is necessary because subscribers are moving, and their radio waves' arrival time at the base station depends on their changing distance from the base station. The greater the distance, the more delay in the signal's arrival time.

The received burst is used by the mobile telephone to determine when its transmission burst should start. The GSM system has some dedicated protection from transmission burst overlap. Each transmit burst has a dedicated guard time of 8.25 bits (30 μsec). This allows mobile devices to operate anywhere in a cell within a distance from the cell site of approximately 4.5 km before overlap may occur. When the distance of the mobile device exceeds 4.5 km from the cell site, the transmission timing is advanced to ensure the transmit burst does not overlap with other mobile devices that are operating within that cell's radio coverage area. The transmitter timing can be advanced in 1/2 bit steps to a maximum of 237 μsec. This limits the maximum distance a GSM mobile telephone can operate from the cell site to approximately 40 km.

Figure 1.21 shows how the relative transmitter timing in a mobile radio (relative to the received signal) is dynamically adjusted to account for the combined receive and transmit delays as the mobile radio is located at different distances from the base station antenna. In this example, the mobile telephone uses a received burst to determine when its burst transmission should start. As the mobile radio moves away from the tower, the transmission time increase and this causes the transmitted bursts to slip outside its time slot when it is received at the base station (possibly causing overlap to transmissions from other radios.) When the base station receiver detects the change in slot period reception, it sends commands to the mobile telephone to advance its relative transmission time as it moves away from the base station and to be retarded as it moves closer.

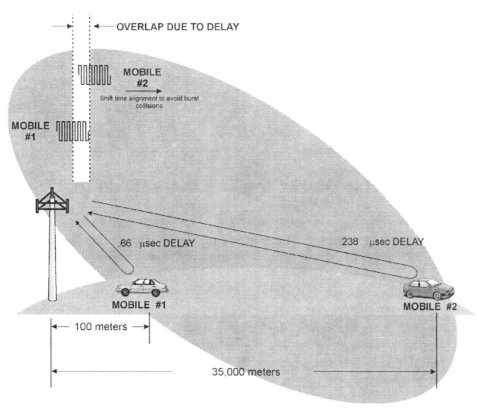

Figure 1.21., Dynamic Time Alignment

It is possible to extend the range of the GSM system beyond the 40 km limit by using extended dynamic time alignment. Extended dynamic time assigns the traffic channel to time slots beyond the 3 time slots offset.

Logical Channels

Logical channels are a portion of a physical communications channel that is used for a particular (logical) communications purpose. The physical channel may be divided in time, frequency or digital coding to provide for these logical channels. The GSM system has two key types of channels; traffic channels and control channels.

Channels can be shared by multiple users (common channels) or they can be used for one-to-one communication (dedicated channels).

Traffic Channels

Traffic channels are the combination of voice and data signals existing within a communication channel.

Traffic Channel or Digital Traffic Channel (TCH or DTC)

A traffic channel is the combination of voice and data signals existing within a communication channel. There are three basic types of traffic channels; full rate, half rate and eighth rate. Variants of these channels also exist.

A full rate traffic channel (TCH/F) dedicates one slot per frame for a communication channel between a user and the cellular system. A half rate traffic channel (TCH/H) dedicates one slot per every two frames for a communication channel between a user and the cellular system. The eighth rate traffic channel (TCH/8) is used only on the SDCCH for exchange of call setup and/or short message service, to provide limited data transmission rates.

Control Channels

Control channels are a communication channel that is used in system (such as a radio control channel) that is dedicated to the sending and/or receiving of controlling messages between devices (such as a base station and a mobile radio). On a mobile radio system, the control channel sends messages that include paging (alerting), access control (channel assignment) and system broadcast information (access parameters and system identification).

Broadcast Channels (BCCH)

Broadcast channels are used to transfer system information such as timing references and synchronization information. The broadcast provides system information, system configuration information (such a paging channel sleep groups), and lists of neighboring radio channels to all mobile devices operating within its radio coverage area.

Each cell site contains a broadcast channel. Mobile devices usually monitor the radio signal strength of cell site broadcast channels to determine which cell site may best provide it with service. The broadcast channel includes a frequency correction channel and a synchronization channel.

Frequency Correction Channel (FCCH)

The frequency correction channel is a signaling channel that provides reference information that allows the mobile device to adjust its frequency so it can better decode the received signals. The frequency correction channel transmission burst occurs before the timing synchronization burst.

Synchronization Channel (SCH)

The synchronization channel is a signaling channel that provides the system timing information that a mobile device needs to adjust its timing so that it can better align, decode, and measure other communication channels.

Cell Broadcast Channel (CBCH)

A cell broadcast channel is an optional channel that carries short messages on the broadcast channel. Each CBCH can transfer about one 80 octet message every 2 seconds [iv]. If the CBCH is included, it shares the same control channel multiframe with the BCCH. This means that CBCH messages can be received in addition to receiving all the BCCH messages.

Common Control Channel (CCCH)

The common control channel is used to coordinate the control of mobile devices operating within its cell radio coverage area. GSM control channels include the random access channel (RACH), paging channel (PCH), and access grant channel (AGCH).

Random Access Channel (RACH)

The random access channel is a signaling control channel that is used by mobile devices to initiate requests for access to the communication system. Responses to service requests that are sent on a RACH channel are provided on the downlink access grant channel (AGCH).

Paging Channel (PCH)

The paging channel is used to send messages (page messages) that alert mobile device of an incoming telephone call (voice call), request for a communicate session (data session), or to request a maintenance service (e.g. location registration update). To alert a mobile device of an incoming call, the paging channel can send a temporary mobile station identity (TMSI) or the international mobile subscriber identity (IMSI).

In addition to sending paging messages, the paging channel is also used to provide information about discontinuous reception (DRx) that allows the mobile device to turn off its circuitry (sleep) during periods between paging groups.

Access Grant Channel (AGCH)

The access grant channel is used to assign a mobile device to a channel where it can begin to communicate with the system. In some cases, the AGCH may assign the mobile device directly to a traffic channel or it may be assigned to an interim control channel where it can communicate with the system before being assigned to a traffic channel.

Random Access Channel (RACH)

The random access channel is a signaling control channel that is used by mobile devices to initiate requests for access to the communication system. Responses to service requests that are sent on a RACH are provided on the access grant channel (AGCH).

Because the distance between the mobile device and the cell site is not typically known when it accesses the system, the access request is attempted using a shortened transmission burst. This prevents potential overlap of the transmission burst with adjacent time slots for the same cell site.

Figure 1.22 shows the basic logical channels used in the GSM system. This diagram shows that the TDMA physical channel is divided into a control channel (time slot 0) and a traffic channel (time slot 4 in this example). The forward logical control channels include the frequency correction channel, synchronization channel, broadcast channel, paging channel, and access grant channel and the reverse logical control channel includes an access request channel. The traffic channel carries user data in both directions. This example shows that while on the traffic channel, fast control channel messages (FACCH) and slow control channel messages (SACCH) can be sent.

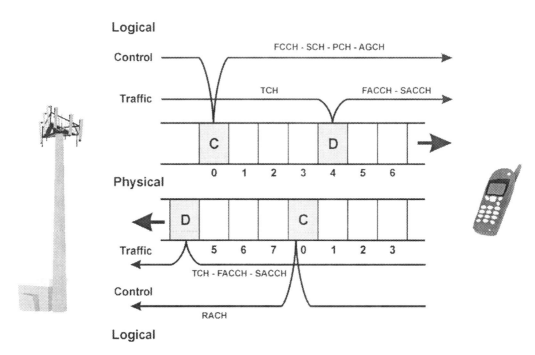

Figure 1.22., Logical Channels Used in GSM Systems

Dedicated Control Channel Signaling

Dedicated control channels are a signaling channel that is used solely for control of a specific device. The GSM system uses dedicated control channels to assist with radio channel assignment and to control the mobile telephone while it is on a traffic channel (voice call or data session).

Stand Alone Dedicated Control Channel (SDCCH)

The stand alone dedicated control channel is a signaling channel that can be used to coordinate the radio channel assignment of a mobile device after it has successfully competed for access. The SDCCH channel is used for off air call setup (OACSU) to allow the mobile device to authenticate and complete other control processes without being assigned to a dedicated traffic channel.

Traffic Channel Signaling

Signaling on the traffic channel is divided into two channels; the Fast Associated Control Channel (FACCH) and the Slow Associated Control Channel (SACCH). The FACCH replaces speech with signal data. The SACCH uses dedicated (scheduled) frames within each burst.

Slow Associated Control Channel (SACCH)

Slow associated control channel (SACCH) is used to continuously transmit certain call processing and control signals at a low bit rate. The SACCH is normally sent along with user data so it does not subtract or use bits from the user data portion. It is therefore sometimes called "out of band" transmission. In full-rate GSM systems, the SACCH data is transmitted in the same time slot that would otherwise be used for digital subscriber traffic. During a scheduled sequence of 26 transmission frames, 24 of these carry digital subscriber traffic, one carries SACCH data, and one is not used.

SACCH is primarily used to transfer radio channel signal quality information from the mobile device to the base station to assist with the handover process. Because SACCH messages do not replace user data (voice signals), the sending of SACCH messages does not affect the quality of speech. However, the data transmission rate of the SACCH is very low and the transmission delay is approximately ½ second.

Figure 1.23 illustrates the SACCH signaling process. This example shows that SACCH messages do not replace voice data, it is sent on a dedicated SACCH time slot on 26 traffic multiframes. Because the SACCH message is distributed over multiple time slots, each SACCH message experiences a delay of approximately 480 msec.

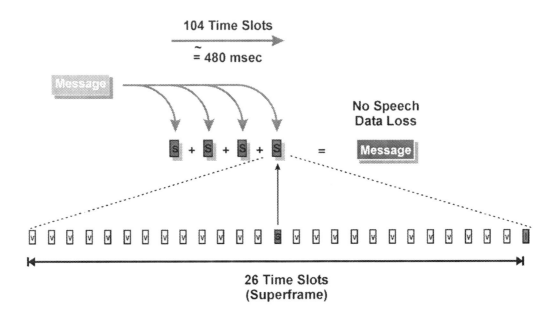

104 Time Slots

$\tilde{}$
= 480 msec

No Speech
Data Loss

26 Time Slots
(Superframe)

Figure 1.23., SACCH Signalin

Fast Associated Control Channel (FACCH)

Fast associated control channel (FACCH) is a logical channel on a digital traffic channel that is typically used to send urgent signaling control messages (such as a handoff or power control message). The FACCH sends messages by replacing speech data with signaling data for short periods of time. In GSM two special reserved bits are used to inform the receiving device if the data in the current time slot is digitally coded subscriber traffic or alternatively a FACCH message. FACCH messaging is also called "in band" signaling. FACCH messages are transmitted over 8 sequential channel bursts.

The sending of FACCH messages replaces user data (usually voice information) and this can degrade speech quality. Because the losses of audio are for very brief periods and the sounds humans' produce does not rapidly change, the speech frames lost due to FACCH messages can be recreated using previous good (successfully received) speech frames. When this occurs, the user hears a brief repeat of a speech sound (for 20 milliseconds) in place of a silence period or noise distortion.

Figure 1.24 shows the basic GSM FACCH signaling process. This diagram shows that the FACCH data replaces (discards) speech frames and replaces them with the FACCH control message data. Each FACCH message is transmitted over 8 sequential channel bursts. The speech information that would normally be transmitted during FACCH transmission is discarded.

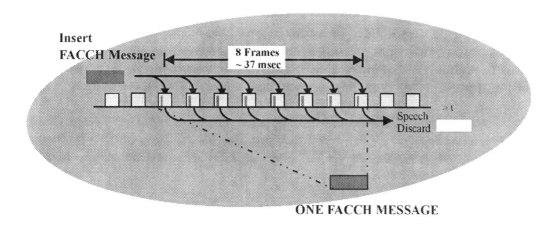

Figure 1.24., FACCH Signaling

DTMF Signaling

Dual tone multifrequency (DTMF) signaling is a means of transferring information from a user to the telephone network through the use of in-band audio tones. Each digit of information is assigned a simultaneous combination of one of a lower group of frequencies and one of a higher group of frequencies to represent each digit or character. There are 8 tones that are capable of producing 16 combinations; 0-9, *, #, A-D. The letters A-D are normally used for non-traditional systems (such as the military telephone systems).

DTMF tones are commonly used to control telephone accessories such as voice mailboxes and interactive voice response (IVR) systems. Unfortunately, DTMF signals are distorted as they pass through the speech coder. The GSM speech coder adds "twist" distortion that changes the relative amplitude of the tone components of the DTMF signal.

Sending DTMF tones through radio channels can also have other impairments. Burst errors that occur during transmission can result in momentary gaps in audio. A momentary gap in audio could result in the false interpretation that at DTMF tone has stopped and a new DTMF has started.

To overcome these challenges, a DTMF tone generator is used into the GSM system to recreate DTMF tones from key presses at the mobile telephone. DTMF signals are sent as DTMF start and DTMF stop messages rather than actual tones.

Figure 1.25 shows how DTMF signals are sent through a GSM system. This example shows that when a user presses the #2 key, a DTMF start message is transmitted. When this message is received at the GSM system, a DTMF #2 tone is created by a DTMF generator. When the user releases the key, a DTMF end message is transmitted. This turns off the DTMF generator in the GSM system.

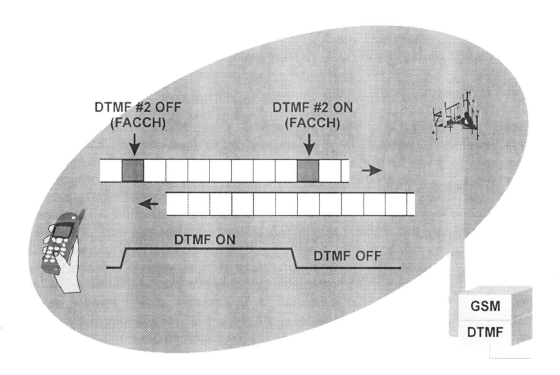

Figure 1.25., DTMF Signaling

GSM Network

GSM networks consist of cell site radio towers, communication links, switching center(s) and network databases and link to public telephone and data networks.

The main switching system in the GSM wireless network is the mobile switching center (MSC). The MSC coordinates the overall allocation and routing of calls throughout the wireless system. Inter-system connections can link different wireless network systems to allow wireless telephones to move from cell site to cell site and system to system. The GSM system defines inter-system connections in detail to allow universal and uniform service availability GSM wireless devices.

The GSM system can be divided into a base station subsystem (BSS), a network and switching system (NSS), and an operation and maintenance subsystem (OMS). The radio parts of the GSM network are contained in the BSS. The switching, databases, and interconnection parts are contained in the NSS. The OMS contains the necessary system to monitor and diagnose system operation.

Figure 1.26 shows a simplified functional diagram of a GSM network. This diagram shows that the GSM system provides for packet data, circuit data, and voice services. Medium-speed circuit data services (up to 56 kbps) are provided by combining multiple coded channels on a single GSM radio channel. This diagram shows that the Base Station (BTS) contains a radio transceiver (radio and transmitter) that converts the radio signal into a data signal (data and digital voice) that can transfer through the network. The BTS is connected to a base station controller (BSC) that coordinates the radio channel assignments. In this example (there are other possible configurations), the packet data at the BSC is routed to a serving GPRS service node (SGSN) and the SGSN is connected to a gateway GPRS service node (GGSN). The circuit switched digital data at the BSC is routed to the mobile switching center (MSC) for connection to the public telephone network and to a data network through an inter-working function (IWF).

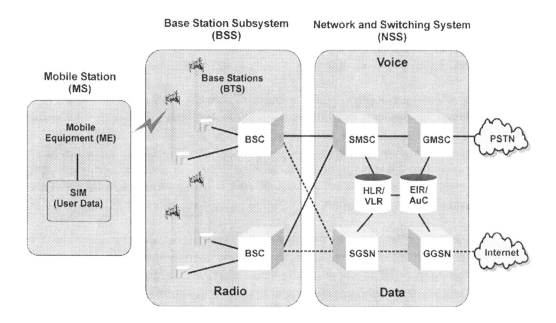

Figure 1.26., GSM Network Parts

Base Stations

Base stations may be stand alone transmission systems that are part of a cell site and is composed of an antenna system (typically a radio tower), building, and base station radio equipment. Base station radio equipment consists of RF equipment (transceivers and antenna interface equipment), controllers, and power supplies. Node B transceivers have many of the same functional elements as a wireless telephone. However, base station radios are coordinated by the GSM system's BSC and have many additional functions than a mobile telephone.

The radio transceiver section is divided into transmitter and receiver assemblies. The transmitter section converts a voice signal to RF for transmission to wireless telephones and the receiver section converts RF from the wireless telephone to voice signals routed to the MSC or packet switching network. The controller section commands insertion and extraction of signaling information.

Unlike the wireless telephone, the transmit, receive, and control sections of a Node B base station are usually grouped into equipment racks. For example, a single equipment rack may contain all of the RF amplifiers or voice channel cards. Unlike analog or early-version digital cellular systems that dedicated one transceiver in each base station for a control channel, the UMTS system combines control channels and voice channels are mixed on a single physical radio channel.

The components of a base station include transceiver assemblies, usually mounted in an equipment rack, each containing multiple assemblies or modules, one for each 5 MHz RF channel. Base station components include the voice or data cards (sometimes called line cards), radio transmitters and receivers, power supplies, and antenna assemblies.

Analog base stations are equipped with a radio channel scanning receiver (sometimes called a locating receiver) to measure wireless telephones' signal strength and channel quality during handover. The UMTS handover process has the advantage of using both base station receiver channel monitoring and radio channel quality information provided back to the system by the wireless telephones (radio signal strength and bit error rate) to assist in the handover process. This information greatly improves the RNCs hand-off decisions.

Radio Antenna Towers

Wireless base station antenna heights can vary from a few feet to more than three hundred feet. Radio towers raise the height of antennas to provide greater area coverage. There may be several different antenna systems mounted on the same radio tower. These other antennas may be used for paging systems, a point to point microwave communication link, or land mobile radio (LMR) dispatch systems. Shared use of towers by different types of radio systems in this way is very common, due to the economies realized by sharing the cost of the tower and shelter. However, great care must be taken in the installation and testing to avoid mutual radio interference between the various systems.

A typical cell site antenna system has multiple antennas. One antenna is used for transmitting and two are used for reception for each radio coverage sector. In some cases, where space or other limitations prevent the use of three separate antennas, two antennas may be used, with one of the two serving as both a transmitter and a receiving antenna, and the other as a receiving antenna only. Special radio frequency filters are used with the shared antenna to prevent the strong transmit signal from causing deleterious effects on the receiver.

The basic antenna options are monopole mount, guy wire, free standing, or man made structures such as water towers, office buildings, and church steeples. Monopole heights range from 30-70 feet; free standing towers range from 20-100 feet; and guy wire towers can exceed 300 feet. Cell site radio antennas can also be disguised to fit in with the surroundings.

Radio Equipment

A radio transmitter in the base station contains audio processing, modulation, and RF power amplifier assemblies. An audio processing section converts digital audio signals from the communications link to channel coded and phase shift modulated signals. The transmitter audio section also inserts control information such as power control messages to the wireless telephone. A modulation section converts the audio signals into proportional phase shifts at the carrier frequency. The RF power amplifier boosts the signal too much higher power levels. This is typically several Watts per RF communication channel compared to the low power of the wireless telephone (typically much less than 1 Watt).

In the UMTS system, the transmitter power level for the control channel is usually fixed to define the cell boundaries (e.g. a control channel). The power level of dedicated (individual) channels may dynamically change to the lowest level possible that allows quality communication with the wireless telephone. This reduction in energy level reduces the overall interference to other wireless telephones that are operating in neighboring cells.

Communication Links

Communication links carry both data and voice information between the MSC, BSCs and the base stations. Options for the physical connections include wire, microwave, or fiber optic links. Alternate communication links are sometimes provided to prevent a single communication link failure from disabling communication [v]. Some terrain conditions may prohibit the use of one type of communication link. For example, microwave systems are not usually used in extremely earthquake-prone areas because they require precise line-of-sight connection. Small shifts in the earth can miss-align microwave transceivers to break communications.

Regardless of the physical type of communication link, the channel format is usually the same. Communication links are typically digital time-multiplexed to increase the efficiency of the communication line. The standard format for time-multiplexing communication channels between cell sites in North America is the 24 channel T1 line, or multiple T1 channels. The standard format outside of North America is the 32 channel (30 useable channels) E1 line.

Repeaters

Repeaters are radio amplifiers that can be used to extend or redirect radio coverage. Repeaters can be a simple remote amplifier or an intelligent frequency translating device. Repeaters are located within the radio coverage area of another base station. The repeater amplifies desired incoming signals and retransmits the signal. The advantage is that the antennas can be located at a longer distance from the base station and it can also provide a greater link budget advantage. This translates into greater coverage and fewer interconnected cell sites required.

Simple repeaters simply boost (amplify) the radio signal in both direction. More advanced frequency translating repeaters receive and decode an RF signal, process the information (changing a few bits), and retransmit the signal on a new frequency. For simple amplifying repeaters, it is important to separate the radio sensing part from the retransmission part to avoid feedback from the repeater's transmitter into its own receiver.

While repeaters are used in GSM systems, there is a maximum distance (maximum number of hops) that repeater signals can be retransmitted. This is due to the transmission time and the maximum time delay. To get extended range from GSM repeaters, extended dynamic time alignment can be used. Extended dynamic time alignment allows the maximum time delay to exceed 3 time slots.

Switching Centers

A switching center coordinates all communication channels and processes. There are two types of switches used in the GSM system; a mobile switching center (MSC) and a packet switching system.

The switching assembly connects the base stations and other networks such as the PSTN or the Internet with either a physical connection (analog) or a logical path (digital). Early analog switches required a physical connection between switch paths. Today's switches use digital switching assemblies that are high-speed matrix memory storage and retrieval systems. These systems provide connections between incoming and outgoing communication lines.

Mobile Switching Centre (MSC)

The mobile switching centre (MSC) processes requests for service from mobile devices and land line callers, and routes calls between the base stations and the public switched telephone network (PSTN). The MSC receives the dialed digits, creates and interprets call processing tones, and routes the call paths.

The basic components of an MSC include system and communication controllers, switching assembly, operator terminals, primary and backup power supplies, wireless telephone database registers, and, in some cases, an authentication (subscriber validation) center.

A system controller coordinates the MSC's operations. A communications controller adapts voice signals and controls the communication links. The switching assembly connects the links between the base station and PSTN. Operator terminals are used to enter commands and display system information. Power supplies and backup energy sources power the equipment. Subscriber databases include a Home Location Register (HLR), used to track home wireless telephones, and a Visitor Location Register (VLR) for wireless telephones temporarily visiting or permanently operating in the system. The authentication center (AC) stores and processes secret keys required to validate the identity of wireless telephones.

The GSM system defines two types of MSC; the serving mobile switching center (SMSC) and the gateway mobile switching center (GMSC). This is the logical separation of the MSC for directly controlling the mobile telephone and providing a bridge between other networks.

The serving mobile switching center (SMSC) is the switch that is connected to the RNC that is providing service directly to the mobile telephone. The SMSC is responsible for coordinating the transfer of calls between different BSC's. When the call is transferred to BSC's that are connected to a different MSC, the role of SMSC will be transferred to the new MSC.

The gateway MSC (GMSC) is the point where the GSM network is connected to the public circuit switched networks (typically the PSTN). All PSTN call connections must enter or leave through the GMSC. The GMSC maintains communication with the SMSC as calls are transferred from one system (or different MSCs within a system) to another.

Serving General Packet Radio Service Support Node (SGSN)

A serving general packet radio service support node is a switching node that coordinates the operation of packet radios that are operating within its service coverage range. The SGSN operates in a similar process of a MSC and a VLR, except the SGSN performs packet switching instead of circuit switching. The SGSN registers and maintains a list of active packet data radios in its network and coordinates the packet transfer between the mobile radios.

Gateway GPRS Support Node (GGSN)

A gateway GPRS support node is a packet switching system that is used to connect a GPRS packet data communication network to other packet networks such as the Internet.

Network Databases

There are many network databases in the GSM network. Some of the key network databases include a master subscriber database (home location register), temporary active user subscriber database (visitor location register), unauthorized or suspect user database (equipment identity register), billing database, and authorization and validation center (authentication).

Home Location Register (HLR)

The home location register (HLR) is a subscriber database containing each customer's international mobile subscriber identity (IMSI) and international mobile equipment identifier (IMEI) to uniquely identify each customer. There is usually only one HLR for each carrier even though each carrier may have many MSCs.

The HLR holds each customer's user profile which includes the selected long distance carrier, calling restrictions, service fee charge rates, and other selected network options. The subscriber can change and store the changes for some feature options in the HLR (such as call forwarding). The MSC system controller uses this information to authorize system access and process individual call billing.

The HLR is a magnetic storage device for a computer (commonly called a hard disk). Subscriber databases are critical, so they are usually regularly backed up, typically on tape or CDROM, to restore the information if the HLR system fails.

Visitor Location Register (VLR)

The visitor location register (VLR) contains a subset of a subscriber's HLR information for use while a mobile telephone is active on a particular MSC. The VLR holds both visiting and home customer's information. The VLR eliminates the need for the MSC to continually check with the mobile telephone's HLR each time access is attempted. The user's required HLR information is temporarily stored in the VLR memory, and then erased either when the wireless telephone registers with another MSC or in another system or after a specified period of inactivity.

Equipment Identity Register (EIR)

The equipment identity register is a database that contains the identity of telecommunications devices (such as wireless telephones) and the status of these devices in the network (such as authorized or not-authorized). The EIR is primarily used to identify wireless telephones that may have been stolen or have questionable usage patterns that may indicate fraudulent use. The EIR has three types of lists; white, black and gray. The white list holds known good IMEIs. The black list holds invalid (barred) IMEIs. The gray list holds IMEIs that may be suspect for fraud or are being tested for validation.

Introduction to GSM

Billing Center (BC)

A separate database, called the billing center, keeps records on billing. The billing center receives individual call records from MSCs and other network equipment. The switching records (connection and data transfer records) are converted into call detail records (CDRs) that hold the time, type of service, connection points, and other details about the network usage that is associated with a specific user identification code. The format of these CDRs is transferred account procedure (TAP) format. The TAP format CDR has evolved into the flexible TAP3 system. The TAP3 system (3rd generation TAP protocol) includes flexible billing record formats for voice and data usage. These billing records are then transferred via tape or data link to a separate computer typically by electronic data interchange (EDI) to a billing system or company that can settle bills between different service providers (a clearinghouse company).

Authentication Center (AuC)

The Authentication Center (AuC) stores and processes information that is required to validate the identity ("authenticate") of a wireless telephone before service is provided. During the authentication procedure, the AC processes information from the wireless telephone (e.g. IMSI, secret keys) along with a random number that is also used by the mobile telephone to produce an authentication response. The AuC compares its authentication response results to the authentication response received from the mobile telephone. If the processed information matches, the wireless telephone passes.

SMS Service Center (SC)

The SMS service center (SC) receives, stores, delivers, and confirms receipt of short messages.

Group Call Register (GCR)

A group call register is a network database that holds a list of group members and the attributes that allow the set-up and processing of calls to and from group members. The GCR holds the membership lists, account features, priority authorization and the current location of group members.

Authentication, Authorization, and Accounting (AAA)

Authentication, authorization, and accounting are the processes used in validating the claimed identity of an end user or a device, such as a host, server, switch, or router in a communication network. Authorization is the act of granting access rights to a user, groups of users, system, or a process. Accounting is the method to establish who, or what, performed a certain action, such as tracking user connection and logging system users.

Wireless Network System Interconnection

System interconnection involves the connection of user data and signaling control messages between different systems. GSM systems are typically connected to public switched telephone network (PSTN) and data networks such as the Internet.

Subscribers can only visit different wireless systems (ROAM) if the systems communicate with each other to handover between systems, verify Roamers, automatically deliver calls, and operate features uniformly. Fortunately, cellular systems can use standard protocols to directly communicate with each other. These inter-system communications use brief packets of data sent via the X.25 packet data network (PDN) or the SS7 PSTN signaling network. SS7 and X.25 are essentially private data communication networks. SS7, which is used by the telephone companies, is available only to telephone companies for direct routing using telephone numbers. The X.25 network does not route directly using telephone numbers. Some MSCs also use other proprietary data connections. No voice information is sent on the SS7 or X.25 networks. Only inter-system signaling such as SS7 mobile applications part (MAP) and intersystem signaling standard 41 (IS-41) is sent between networks to establish, authenticate and maintain communication paths.

Ideally, inter-system signaling is independent of cellular network radio technology, but this can be difficult between systems where radio technologies differ. Consider inter-system hand-off between a GSM capable and a UMTS capable cell site (assuming the wireless telephone were capable of both). The UMTS system uses soft handover while GSM does not. As new features in wireless networks are added, inter-system signaling messages, standards that define them, and equipment that processes them must change.

Communication between MSCs is performed either by a proprietary or standard protocol. Standard protocols such as SS7 mobile applications part (MAP) or IS-41 allow MSCs of different makes to communicate with few or no changes to the MSC. Regardless of whether a standard (e.g. IS-41) protocol or a manufacturers private (proprietary) protocol is used, the underlying data transferred via inter-system signaling is the same. If changes are required to communicate with a different protocol, an interface (protocol converter) changes the proprietary protocol to standard protocol. The interface has a buffer that temporarily stores data elements being sent by the MSC and reformats it to the SS7 MAP or IS-41 protocol. Another buffer stores data until it can be sent via the control signaling network.

Public Switched Telephone Network (PSTN)

Public switched telephone networks are communication systems that are available for public to allow users to interconnect communication devices. Public telephone networks within countries and regions are standard integrated systems of transmission and switching facilities, signaling processors, and associated operations support systems that allow communication devices to communicate with each other when they operate.

Public Packet Data Network (PPDN)

A packet data network that is generally available for commercial users (the public). An example of a PPDN is the Internet.

Interworking Function (IWF)

Interworking functions are systems and/or processes that attach to a communications network that is used to process and adapt information between dissimilar types of network systems. IWFs in the GSM system may include data gateways that convert circuit switched data from the MSC to the Internet.

Customized Applications for Mobile Network Enhanced Logic (CAMEL)

Customized applications for mobile network enhanced logic is an intelligent network service specification that allows service providers to create custom service applications for mobile telephone systems. CAMEL operates on a "services creation node" in a GSM or WCDMA network. Examples of CAMEL applications include time of day call forwarding, multiple telephone extension service, and automatic call initiation on special conditions (trigger).

Device Addressing

GSM mobile terminals are uniquely identified by the mobile identification number through the use of mobile station ISDN (MSISDN), of International Mobile Subscriber Identity (IMSI), International Mobile Equipment Identifier (IMEI), and a temporary mobile station identity (TMSI). In addition temporary IP addresses may be assigned, as required.

Mobile Station ISDN (MSISDN)

The mobile station ISDN is the phone number assigned to mobile telephones. This number is compatible with the E.164 international public telephone numbering plan.

International Mobile Subscriber Identity (IMSI)

The international mobile subscriber identity is an identification number for that is assigned by a mobile system operator to uniquely identify a mobile telephone.

International Mobile Equipment Identifier (IMEI)

An International Mobile Equipment Identifier (IMEI) is an electronic serial number that is contained in a GSM mobile radio. The IMEI is composed of 14 digits. Six digits are used for the type approval code (TAC), two digits are used for the final assembly code (FAC), six digits are used for the serial number and two digits are used for the software version number.

Temporary Mobile Station Identity (TMSI)

A temporary mobile station identity (TMSI) is a number that is used to temporarily identify a mobile device that is operating in a local system. A TMSI is typically assigned to a mobile device by the system during its' initial registration. The TMSI is used instead of the International Mobile Subscriber Identity (IMSI) or the mobile directory number (MDN). TMSIs may be used to provide increased privacy (keeping the telephone number private) and to reduce the number of bits that are sent on the paging channel (the number of bits for a TMSI are much lower than the number of bits that represent an IMSI or MDN).

GSM System Operation

There are many other processes a mobile telephone must perform to operate in a GSM network. The basic call processing operation of a mobile telephone includes initialization, call origination, call reception (paging), and handover.

When a subscriber unit is first powered on, it initializes by scanning for a control channel and tuning to the strongest one it finds. During initialization, it acquires all of the system information needed to monitor for paging messages and information about how to access the system. After initialization, the subscriber unit enters idle (sleep and wake cycle) mode and waits either to be paged for an incoming call or for the user to place a call (access). When a call is to be received or placed, the subscriber unit enters system access mode to try to access the system via a control channel. When access is granted, the control channel commands the subscriber unit to tune to a digital traffic channel. The subscriber unit tunes to the designated channel, and enters conversation mode. As the subscriber unit moves out of range of one cell site radio coverage area, it is handed over to a radio traffic channel at another nearby cell site.

Mobile Telephone Initialization

Mobile telephone initialization is the process of a mobile device searching for a system broadcast radio channel, synchronizing with the system and obtaining system parameters that it will use to coordinate its access to the system.

Initialization phase begins when the mobile device is first turned on. It initially looks to the subscriber identity module (SIM) card for a preferred control channel list. If there is no list, the mobile device scans all of the available radio channels to find a control channel.

Broadcast channel information includes system identification and access control information (such as access priorities and initial transmit power levels). A mobile telephone may find several broadcast radio channels that have acceptable signal strength levels. It may be programmed to use the radio channel with the highest signal level or it may prefer to use radio channels that are from its preferred system (e.g. home system).

Updating Location (Registration)

Location registration is the process that is used by mobile devices to inform the wireless system of their location and availability to receive communications services (such as incoming calls). The reception of registration requests allows a wireless system to route incoming messages to the radio base station or transmitter where the mobile device has recently registered.

The process of registration is typically continuous. Mobile devices register when they power on, when they move between new radio coverage areas, when requested by the system, and when the mobile device is power off.

Because the registration process consumes resources of the system (channel capacity and system servicing capacity), there is a tradeoff between regularly maintaining registration information and the capacity of the system. During periods of high system usage activity, registration processes may be reduced.

Waiting for Calls (Idle)

During idle mode, the mobile device monitors the control channel to update system access parameters, to determine if it has been paged or received an order, or to initiate a call (if the user is placing a call) or to start a data session (if the user has started a data application).

If the mobile device has discontinuous reception (sleep mode) capability, and if the system supports it, the mobile device turns off its receiver and other non-essential circuitry for a fixed number of burst periods. The system knows that it has commanded the mobile device to sleep, so it does not send pages designated for that mobile device during the sleep period. Because control channels are on only one of the 8 time slots in a frame, the mobile device can scan neighboring control channels during the unused time slots. If a better control channel (higher signal strength or better bit error rate) is available, the mobile device may retune to the new channel frequency.

The mobile device then monitors the paging control channel to determine if it has received a page. If a call is to be received, an internal flag is set indicating that the mobile device is entering access mode in response to a page. If the system sends an order such as a registration message, an internal flag is set indicating that the mobile device is attempting access in response to an order. When a user initiates a call, an internal flag is set indicating that the access attempt is a call origination, and dialed digits will follow the access request.

System Access

System access control is the process of gaining the attention of the system, obtaining authorization to use system services, and the initial assignment to the communication channel to setup a communication session.

Access control and initial assignment occurs when a mobile device responds to a page (incoming connection request), desires to setup a call, or any attempt by the mobile device. Access to the GSM system is a random occurrence (not usually preplanned.) To avoid access "collisions" between mobile devices, a seizure collision avoidance process is used. Before a mobile device attempts access to the system, it first waits until the channel is available (not busy serving other users). The mobile device then begins transmitting an access request message on the random access channel (RACH) at a power level assigned by the broadcast channel.

The access request message uses a shortened burst that has a predefined sequence that allows the system to easily detect that an access request message has been sent. Using a shortened burst (87 bits) also prevents the potential of overlapping the transmission burst with adjacent time slots when the cell size is larger than the guard period allows (4.5 km).

The access competition process includes sending a request sent on the RACH (in a particular slot), listening to the system to determine if it has recognized and responded to the access request, waiting a random amount of time before sending another access request if the system has not responded, and limiting the maximum number of access attempts to a value the mobile device has received from the system.

To uniquely identify access request messages, the mobile device creates and sends a 5-bit random number inside the access request. The mobile device reviews the access response message to determine if the system is responding to its access request message or if it is responding to the access request of another mobile device.

If an access request message does not gain the attention of the system in an expected period of time, the mobile device will wait a random amount of time and retransmit an access request message. Each time this process repeats, the mobile device will increase in the random wait time until the mobile device has reached the maximum number of system access requests allowed by the system.

If the system acknowledges the mobile device's request for service, the mobile device will send additional information to the system that allows it to setup a dedicated communications channel where conversation or data transmission can begin. The dedicated communication channel may be a traffic channel (for user voice and data) or an interim signaling channel, the stand alone dedicated control channel (SDCCH), which allows the mobile device to setup (e.g. authenticate) while it is waiting for a traffic channel assignment.

Mobile Call Origination

Mobile call origination is the process of initiating a communication session by a mobile device. Mobile origination typically occurs when a user dials a telephone number and presses the SEND button.

When initiating a call, a mobile telephone attempts to gain service from the GSM system by transmitting a system access request and indicating the access request is a call origination type. The access type is indicated by a 3-bit code that is contained in the access request message.

Figure 1.27 shows a functional diagram of how a mobile telephone initiates a call to a GSM network. In step 1, the mobile telephone sends system access requests message indicating it desires to initiate a call. When the system acknowledges the request, the mobile telephone is assigned to a traffic channel (step 2). The dialed digits are then sent to the system and the call is routed to the destination telephone (step 3). If the called person answers, the GSM system will open an audio path between the GSM user and the destination telephone (step 4).

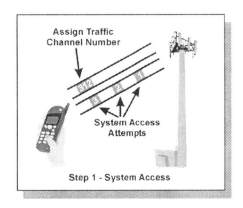

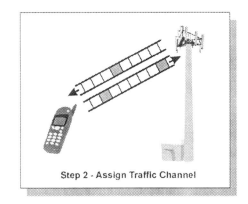

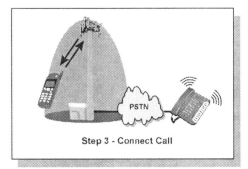

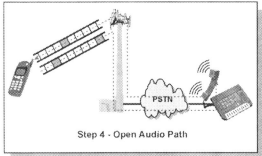

Figure 1.27., Mobile Call Origination

Transferring Calls Between Cell Sites (Handover)

Handover is a process where a mobile device that is operating on a particular channel is reassigned to a new channel. This can be a new frequency channel, new code channel, or a new logical channel. The handoff process is often used to allow subscribers to travel throughout the large radio system coverage area by switching the calls (handoff) from cell-to-cell (and different channels) with better coverage for that particular area when poor quality conversation is detected.

Handoff may also occur when a mobile device requests a service that can only be provided by a radio channel that has different service capabilities. This might mean assignment from a GSM traffic (voice) channel to a GPRS packet data channel.

The GSM system can handover a communication channel when the base station controller (BSC) determines the radio channel quality between the mobile device and the base station has fallen below an acceptable level and a better radio channel is available for call transfer. The BSC can use information that it receives from the mobile device to assist in its' handover decisions. The mobile device can measure the received signal strength (RSSI) and bit error rate (BER) and return this information to the BSC via the base station.

The frame structure of the GSM radio channel allows mobile telephones to monitor nearby base station radio channels during idle time slots. The GSM system provides the mobile telephone with a list of nearby cell site radio channels on the broadcast control channel. The mobile device provides channel quality information back to the system on the list of candidate radio channels. When the GSM system determines that the mobile telephone signal quality has fallen below a desired level and it has available radio channels in the area, a handover message is sent to the mobile telephone. This commands it to tune to a new radio channel.

During the handoff command, the system may send a shortened transmission bursts to avoid potential collisions due to unknown distance and time alignment. The use of a shortened burst means that some additional user data information will be lost during the transfer. To avoid using a shortened burst during handoff, the mobile telephone may pre-time synchronize to the new radio channel. This allows the mobile telephone to immediately transmit normal audio bursts during the handover process.

Receiving a Call on a Mobile Telephone

The GSM system sends paging messages to alert mobile devices that they are receiving a call, command, or message. Mobile devices listen for paging messages with their identification code (IMSI number or TMSI) on a paging channel.

After a mobile device has registered with the system, it is assigned to a paging group. The paging group is identified by a paging indicator (PI) that is provided at the beginning of the paging message group. The mobile device first reads the PI to determine if it should remain awake to receive the paging group or if it can go to sleep as its identification code is not associated with the particular paging group.

To receive a call, the mobile device synchronizes to the system and continuously monitors the paging channel. When the mobile device receives its identification number on the paging channel, it will attempt to access the mobile system indicating its access request is in response to a paging message. The system may then validate the identity of the mobile device and assign it to a traffic channel. The mobile device then alerts the user of an incoming call (ringing the mobile device) and if the user answers the call (pressing SEND), the mobile device alerts the system the call has been answered and the cellular system can connect the audio path between the mobile device and the caller.

Figure 1.28 shows the basic process for receiving calls on a GSM system. In step 1, the mobile telephone receives a page message in its paging group. The mobile telephone sends access request messages to the system indicating the access is in response to a page message (step 2). The system assigns the mobile telephone to a traffic channel (step 3). The audio paths are then opened between the caller and the mobile telephone (step 4).

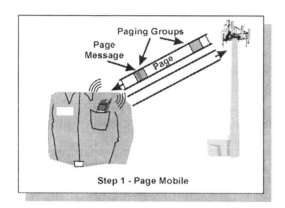
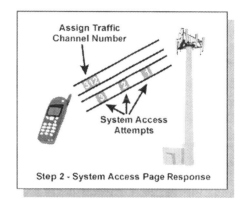
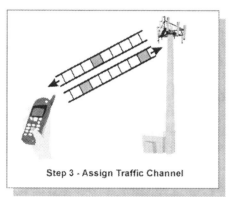
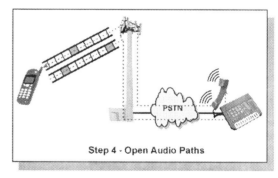

Figure 1.28., Receiving a Call on a Mobile

Conversation Mode

Conversation mode is the process of managing the communication session when a mobile device is transferring voice signals to and from a base station. When in the conversation mode, the base transceiver station (BTS) continuously controls the mobile device during the communication session. These control tasks include power level control, handover, alerting, etc. The base station exercises control during the communication session through signaling on control channels.

To enter the conversation mode, the base station must open a communication channel with the mobile device. When the connection is opened, each frame or packet that is received by the base station can be transferred to the assigned communication line or channel on a multichannel communication line.

When a communication session is complete (e.g. the user presses end or closes their email or web browsing application), the connection is closed and the base station may assign other users to the radio resources.

Connected Mode

Connected mode is the process of managing the communication session when a mobile device is transferring packet data signals to and from a base station. When in the connected mode, the base transceiver station (BTS) continuously controls the mobile device during the communication session. These control tasks include power level control, handover, alerting, etc. The base station exercises control during the communication session through signaling on control channels.

During the connected mode, communication session processing tasks include the insertion and extraction of control messages that allow functions such as power control monitoring and control, handover operation, adding or terminating additional communication sessions (logical channels), and other mobile device operational functions.

When sending packet data, the mobile device does not have to continuously transmit. When in the data communication mode, the base station associates (maps) the radio link to a data connection or to an IP address (for IP routing through networks). When the connected mode is used for data transmission such as web browsing, the typical data transmission activity is less than 10%. Other mobile devices can use the channels during inactive data transmission periods allowing a system to serve many (hundreds) simultaneous data users for each WCDMA radio channel.

Authentication

Authentication is a process of exchanging information between a communications device (typically a user device such as a mobile phone) and a communications network that allows the carrier or network operator to confirm the true identity of the user (or device). This validation of the authenticity of the user or device allows a service provider to deny service to users that cannot be identified. Thus, authentication inhibits fraudulent use of a communication device that does not contain the proper identification information. The GSM system may require the mobile device to authenticate with the system during the system access process.

The authentication algorithm used in the GSM system is contained in the subscriber identity module (SIM) card. The GSM authentication process can use different versions of authentication [iv]. The GSM authentication process starts by the transmission of a random number (RAND) from the base station. This random number is used along with other information including the secret data value (Ki) to calculate a signed result (SRES). The secret number Ki is stored in both the mobile telephone and GSM system and it is not transmitted over the radio link. When the GSM system performs the authentication process, it compares the SRES it calculates to the SRES returned by the mobile telephone. If both SRES match, the GSM system allows call processing to continue. The codes generated in the authentication step may be used for voice privacy (encryption) mode.

Figure 1.29 shows the basic GSM authentication process. As part of a typical authentication process, a random number that changes periodically (RAND) is sent from the base station. This number is regularly received and temporarily stored by the mobile radio. The random number is then processed with the shared secret data (Ki) that has been previously stored in the SIM card along with other information in the subscriber to create an authentication response (SRES). The authentication response is sent back to the system to validate the mobile radio. The system processes the same information to create its own authentication response. If both the authentication responses match, service may be provided. This process avoids sending any secret information over the radio communication channel.

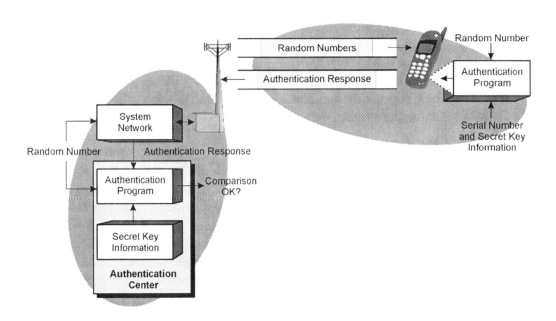

Figure 1.29., GSM Authentication Process

GSM Future Evolution

The evolution of GSM includes packet data transmission rates and high-speed data wideband code division multiple access (WCDMA).

Enhanced Data for Global Evolution (EDGE)

Enhanced data for global evolution is an evolved version of the global system for mobile (GSM) radio channel that uses new phase modulation and packet transmission to provide for advanced high-speed data services. The EDGE system uses 8 level phase shift keying (8PSK) to allow one symbol change to represent 3 bits of information. This is 3 times the amount of information that is transferred by a standard 2 level GMSK signal used by the first generation of GSM system. This results in a radio channel data transmission rate of 604.8 kbps and a net maximum delivered data transmission rate of approximately 474 kbps. The advanced packet transmission control system allows for constantly varying data transmission rates in either direction between mobile radios.

Figure 1.30 shows how a standard GSM radio channel is modified to use a new, more efficient modulation technology to create a high-speed packet data EDGE system. The EDGE system uses either 8 level quadrature phase shift keying (QPSK) modulation or the standard GMSK modulation (used by 2nd generation GSM systems.) This allows EDGE technology to be merged on to existing GSM systems as standard GSM mobile telephones will ignore the EDGE modulated time slots that they cannot demodulate and decode.

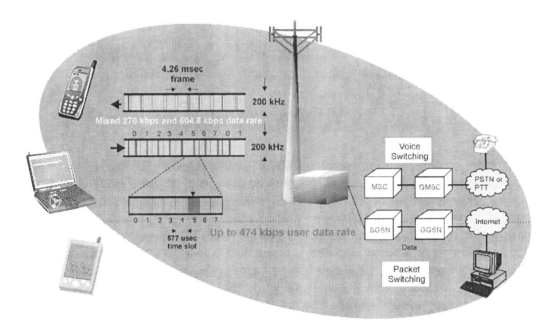

Figure 1.30, EDGE System

Wideband Code Division Multiple Access (WCDMA)

Wideband code division multiple access is a 3rd generation mobile communication system that uses wideband code division multiple access (WCDMA) technology. The WCDMA infrastructure is compatible with GSM mobile radio communication system. WCDMA provides for high-speed data and voice communication services. Installing or upgrading to WCDMA technology allows mobile service providers to offer their customers wireless broadband (high-speed Internet) services and to operate their systems more efficiently (more customers per cell site radio tower).

The WCDMA system is composed of mobile devices (wireless telephones and data communication devices called user equipment - UE), radio towers (cell sites called Node Bs), and an interconnection system (switches and data routers). The WCDMA system uses two types of radio channels; frequency division duplex (FDD) and time division duplex (TDD). The FDD radio channels are primarily used for wide area voice (audio) channels and data services. The TDD channels are typically used for systems that do not have the availability of dual frequency bands.

Figure 1.31 shows a simplified diagram of a WCDMA system. This diagram shows that the WCDMA system includes various types of mobile communication devices (called user equipment - UE) that communicate through base stations (node B) and a mobile switching center (MSC) or data routing networks to connect to other mobile telephones, public telephones, or to the Internet via a core network (CN). This diagram shows that the WCDMA system is compatible with both the 5 MHz wide WCDMA radio channel and the narrow 200 kHz GSM channels. This example also shows that the core network is essentially divided between voice systems (circuit switching) and packet data (packet switching).

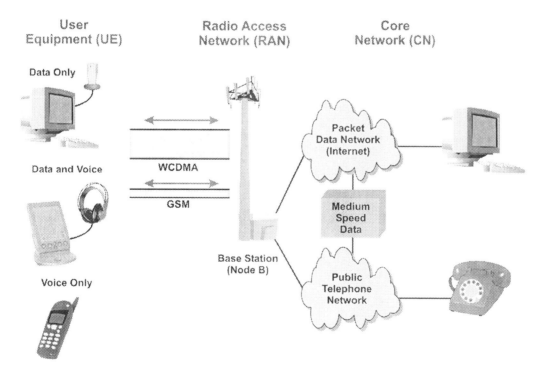

Figure 1.31., WCDMA System

References:

[i]. www.GSMworld.com, 26 October, 2004.

[ii]. D.M. Balston, R.V. Macario, "Cellular Radio Systems", Artech House, 1993, pg. 126.

[iii]. Michel Mouly, Marie-Bernadette Pautet, "The GSM System for Mobile Communications", M. Mouly et Marie-B Pautet, Palaiseau, France, pp. 218-221.

[iv]. Michel Mouly, Marie-Bernadette Pautet, "The GSM System for Mobile Communications", M. Mouly et Marie-B Pautet, Palaiseau, France, pg. 193

[v]. CTIA Winter Exposition, "Disaster Experiences", Reno Nevada, February 6, 1990.

[vi]. Michael Mouly and Marie-Bernadette Pautet, The GSM System for Mobile Communications, (M. Mouly et Marie-B. Pautet), pp 478,479.

Index

Wireless Books
by ALTHOS Publishing

Wireless Systems

ISBN: 0-9728053-4-6   Price: $34.99
Authors: Lawrence Harte, Dave Bowler, Avi Ofrane, Ben Levitan
#Pages 368   Copyright Year: 2004

Wireless Systems; Cellular, PCS, 3G Wireless, LMR, Paging, Mobile Data, WLAN, and Satellite explains how wireless telecommunications systems and services work. There are many different types of wireless systems competing to offer similar types of voice, data, and multimedia services. This book describes what the functional parts of these systems are and the basics of how these systems operate. With this knowledge,

Wireless Dictionary

ISBN: 0-9746943-1-2   Price: $39.99
Author: Althos   #Pages: 628   Copyright Year: 2004

The Wireless Dictionary is the Leading wireless industry resource. The Wireless Dictionary provides definitions and illustrations covering the latest voice, data, and multimedia services and provides the understanding needed to communicate with others in the wireless industry. This book is the perfect solution for those involved or interested in the operation of wireless devices, networks, and service providers.

Introduction to 802.11 Wireless LAN (WLAN)

ISBN: 0-9746943-4-7   Price: $14.99
Author: Lawrence Harte   #Pages: 52   Copyright Year: 2004

Introduction to 802.11 Wireless LAN (WLAN), Technology, Installation, Setup, and Security book explains the functional parts of a Wireless LAN system and their basic operation. You will learn how WLANs can use access points to connect to each other or how they can directly connect between two computers. Explained is the basic operation of WLAN systems and how the performance may vary based on a variety of controllable and uncontrollable events. This book will explain the key differences between the WLAN systems.

Introduction To Wireless Systems

ISBN: 0-9742787-9-3   Price: $11.99
Author: Lawrence Harte,   #Pages: 68   Copyright Year: 2003

Introduction to Wireless Systems book explains the different types of wireless technologies and systems, the basics of how they operate, the different types of wireless voice, data and broadcast services, key commercial systems, and typical revenues/costs of these services. Wireless technologies, systems, and services have dramatically changed over the past 5 years. New technology capabilities and limited restrictions are allowing existing systems to offer new services.

Althos Publishing, 404 Wake Chapel Road, Fuquay NC 27526 USA
1-919-557-2260 1-800-227-9681 Fax 1-919-557-2261 WWW.AlthosBooks.com

Introduction to Paging Systems

ISBN: 0-9746943-7-1   Price: $14.99
Author: Lawrence Harte   #Pages: 48   Copyright Year: 2004

Introduction to Paging Systems describes the different types of paging systems, what services they can provide, and how they are changing to meet new types of uses. This book explains the different types of paging systems and how they are changing. Explained is how and why paging systems are transitioning from one-way systems to two-way systems.

Introduction to Satellite Systems

ISBN: 0-9742787-8-5   Price: $11.99
Author: Ben levitan, Lawrence Harte   #Pages: 48   Copyright Year: 2004

In 2003, the satellite industry was a high-growth business that achieved over $83 billion in annual revenue. This book offers an introduction to existing and soon to be released satellite communication technologies and services. It covers how satellite systems are changing, growth in key satellite markets and key technologies that are used in satellite systems.

Introduction to Mobile Data

ISBN: 0-9746943-9-8   Price: $14.99
Author: Lawrence Harte   #Pages: 628   Copyright Year: 2004

Introduction to Mobile Data explains how people use devices that can send data via wireless connections, what systems are available for providing mobile data service, and the services these systems can offer. This book explains the basics of circuit switched and packet data via wireless mobile systems. Included are descriptions of various public and private systems that are used for data and messaging services.

Introduction to Private Land Mobile Radio

ISBN: 0-9746943-6-3   Price: $14.99
Author: Lawrence Harte   #Pages: 50   Copyright Year: 2004

Introduction to Private Land Mobile Radio explains the different types of private land mobile radio systems, their basic operation, and the services they can provide. This book covers the basics of private land mobile radio systems including traditional dispatch, analog trunked radio, logic trunked radio (LTR), and advanced digital land mobile radio systems. Described are the basics of LMR technologies including simplex,and half-duplex.

Introduction to GSM Systems

ISBN: 1-9328130-4-7   Price: $14.99
Author: Lawrence Harte   #Pages: 48   Copyright Year: 2004

Introduction to GSM describes the fundamental components, key radio and logical channel structures, and the basic operation of the GSM system. This book explains the basic technical components and operation of GSM technology. You will learn the physical radio channel structures of the GSM system along with the basic frame and slot structures.

Althos Publishing, 404 Wake Chapel Road, Fuquay NC 27526 USA
1-919-557-2260 1-800-227-9681 Fax 1-919-557-2261 WWW.AlthosBooks.com

Internet Telephone Basics

ISBN: 0-9728053-03   Price: $29.99
Author: Lawrence Harte   #Pages:226   Copyright Year: 2003

Internet Telephone Basics explains how and why people and companies are changing to Internet Telephone Service. Learn how much money can be saved using Internet telephone service and how you can to use standard telephones and dial the same way. It describes how to activate Internet telephone service instantly and how to display your call details on the web.

Voice Over Data Networks for Managers

ISBN: 0-9728053-2-X   Price: $49.99
Author: Lawrence Harte   #Pages:352   Copyright Year: 2003

Voice over Data Networks for Managers explains how to reduce communication costs 40% to 70%, keep existing telephone systems, and ways to increase revenue from new communication applications. Discover the critical steps companies should take and risks to avoid when transitioning from private telephone systems (KTS, PBX, and CTI) to provide new services.

Patent or Perish

ISBN: 0-9728053-3-8   Price: $39.95
Author: Eric Stasik   #Pages:220   Copyright Year: 2003

Patent or Perish Explains in clear and simple terms the vital role patents play in enabling high technology firms to gain and maintain a competitive edge in the knowledge economy. Patent or Perish is a Guide for Gaining and Maintaining Competitive Advantage in the Knowledge Economy. In a world where knowledge has value and knowledge creates value, ideas are the new source of wealth. This book describes how technologies like the Internet remove traditional barriers to entry and enable competitors to quickly.

Telecom Basics 3rd Edition

ISBN: 0-9728053-5-4   Price: $34.99
Author: Lawrence Harte   #Pages:356   Copyright Year: 2003

This introductory book provides the fundamentals of signal processing, signaling control, can call processing technologies that are used in telecommunication systems. Covered are the key facets of voice and data communications, ranging from such basics as to how a telephone set works to more complex topics as how to send voice over data networks and the ways calls are processed in public and private telephone systems.

Introduction to Cable Television Systems

ISBN: 0-9728053-6-2   Price: $12.99
Author: Lawrence Harte   #Pages: 48   Copyright Year: 2004

Community access television (CATV) is a television distribution system that uses a network of cables to deliver multiple video, data, and audio channels. This excerpted chapter from Telecom Systems provides an overview of cable television systems including cable modems, digital television, high definition television (HDTV), and the market growth of cable television and advanced services such as video on demand.

Althos Publishing, 404 Wake Chapel Road, Fuquay NC 27526 USA
1-919-557-2260 1-800-227-9681 Fax 1-919-557-2261 WWW.AlthosBooks.com

Introduction to Private Telephone Networks 2ⁿᵈ Edition

ISBN: 0-9742787-2-6   Price: $12.99
Author: Lawrence Harte   #Pages: 48   Copyright Year: 2004

Private telephone networks are communication systems that are owned, leased or operated by the companies that use these systems. They primarily allow the interconnection of multiple telephones within the private network with each other and provide for the sharing of telephone lines from a public telephone network.

Introduction to Telecom Billing

ISBN: 0-9742787-4-2   Price: $11.99
Author: Lawrence Harte   #Pages: 36   Copyright Year: 2003

This book explains how companies bill for telephone and data services, information services, and non-communication products and services. Billing and customer care systems convert the bits and bytes of digital information within a network into the money that will be received by the service provider. To accomplish this, these systems provide account activation and tracking, service feature selection, selection of billing rates for specific calls, invoice creation, payment entry and management of communication with the customer.

Introduction to Public Switched Telephone Networks 2ⁿᵈ Edition

ISBN: 0-9742787-6-9   Price: $34.99
Author: Lawrence Harte   #Pages: 48   Copyright Year: 2004

Public telephone networks are unrestricted dialing telephone networks that are available for public use to interconnect communications devices. There are also descriptions of many related topics, including: Local loops, switching systems, numbering plans, market growth, public telephone system interconnections, and common channel signaling (SS7),

Introduction to SS7 & IP Telephony

ISBN: 0-9746943-0-4   Price: $14.99
Author: Lawrence Harte   #Pages: Copyright Year: 2004

The Introduction to Signaling System 7 (SS7) and IP control system that is used in public switched telephone networks (PSTN) can be interconnected to other types of systems and networks using Internet Protocol (IP). Some of the interconnection issues relate to how the control of devices can be performed using dissimilar systems.

Introduction to IP Telephony

ISBN: 0-974278-7-7   Price: $12.99
Author: Lawrence Harte   #Pages: Copyright Year: 2003

This "Introduction to IP Telephony" book explains why companies are converting some or all of their telephone systems from dedicated telephone systems (such as PBX) to more standard IP telephony systems. These conversions allow for telephone bill cost reduction, increased ability to control telephone services, and the addition of new telephone information

Althos Publishing, 404 Wake Chapel Road, Fuquay NC 27526 USA
1-919-557-2260 1-800-227-9681 Fax 1-919-557-2261 WWW.AlthosBooks.com

Signaling System Seven (SS7) Basics 3rd Edition

ISBN: 0-9728053-7-0   Price: $34.99
Author: Lawrence Harte   #Pages: 276   Copyright Year: 2003
This introductory book explains the operation of the signaling system 7 (SS7) and how it controls and interacts with public telephone networks and VoIP systems. SS7 is the standard communication system that is used to control public telephone networks. In addition to voice control, SS7 technology now offers advanced intelligent network features and it has recently been updated to include broadband control capabilities.

Introduction to SIP IP Telephony Systems

ISBN: 0-9728053-8-9   Price: $14.99
Author: Lawrence Harte   #Pages: 117   Copyright Year: 2004
This book explains why people and companies are using SIP equipment and software to efficiently upgrade existing telephone systems, develop their own advanced communications services, and to more easily integrate telephone network with company information systems. This book provides descriptions of the function parts of SIP systems along with the fundamentals of how SIP systems operate.

Telecom Systems

ISBN: 0-9728053-9-7   Price: $34.99
Author: Lawrence Harte   #Pages: 480   Copyright Year: 2004

This book Telecom Systems shows the latest telecommunications technologies are converting traditional telephone and computer networks into cost competitive integrated digital systems with yet undiscovered applications. These systems are continuing to emerge and become more complex.Telecom Systems explains how various telecommunications systems and services work and how they are evolving to meet the needs of bandwidth

Introduction to Transmission Systems

ISBN: 0-9742787-0-X   Price: $14.99
Author: Lawrence Harte   #Pages: 52   Copyright Year: 2004

This book explains the fundamentals of transmission lines and how radio waves, electrical circuits, and optical signals transfer information through a communication medium or channel on carrier signals. It also explains the ways that a single line can be divided into multiple channels and how signals are carried over transmission lines in analog or digital form.

Tehrani's IP Telephony Dictionary

ISBN: 0-9742787-1-8   Price: $39.99
Author: Althos   #Pages: 628   Copyright Year: 2003

Tehrani's IP Telephony Dictionary, The Leading VoIP and Internet Telephony Resource provides over 10,000 of the latest IP Telephony terms and more than 400 illustrations to define and explain latest voice over data network (VoIP) technologies and services. It provides the references needed to communicate with others in the communication industry.

Althos Publishing, 404 Wake Chapel Road, Fuquay NC 27526 USA
1-919-557-2260 1-800-227-9681 Fax 1-919-557-2261 WWW.AlthosBooks.com

Practical Patent Strategies Used by Successful Companies

ISBN: 0-9746943-3-9 Price: $14.99
Author: Eric Stasik #Pages: Copyright Year: 2004

This book explains how companies can use patent strategies to achieve their business goals. Patent strategies may be considered abstract legal or economic concepts. Examining how patents are used by leading companies in specific business applications can provide great insight to their practical use and application in your business plan. This book presents in plain and clear language why having a patent strategy is important.

Introduction to xHTML

ISBN: 0-9328130-0-4 Price: $34.99
Author: Lawrence Harte #Pages: Copyright Year: 2004

This book explains what is xHTML Basic, when to use it, and why it is important to learn. You will discover how the xHTML Basic language was developed and the types of applications that benefit from xHTML Basic programs. The basic programming structure of xHTML Basic is described along with the basic commands including links, images, and special symbols.

Introduction to SS7

ISBN: 1-9328130-2-0 Price: $14.99
Author: Lawrence Harte #Pages: Copyright Year: 2004

This introductory book explains the basic operation of the signaling system 7 (SS7). SS7 is the standard communication system that is used to control public telephone networks. This book will help the reader gain an understanding of SS7 technology, network equipment, and overall operation. It covers the reasons why SS7 exists and is necessary.

Creating RFPs for IP Telephony Communications Systems

ISBN: 1-9328131-1-X Price: $19.99
Author: Lawrence Harte #Pages: Copyright Year: 2004

This book explains the typical objectives and processes that are involved in the creation and response to request for proposals (RFPs) for IP Telephony systems and services. It covers the key objectives for the RFP process, whose involved in the creation and management of the RFP, and how vendors are invited, evaluated, and notified of the RFP vendor selection result. You will learn what are RFPs and RFQs and why and when companies use and RFPs for IP Telephony Systems.

ATM Basics

ISBN: 1-9328131-3-6 Price: $29.99
Author: Lawrence Harte #Pages: Copyright Year: 2004

Asynchronous Transfer Mode (ATM) is a high-speed packet switching network technology industry standard. ATM networks have been deployed because they offer the ability to transport voice, data, and video signals over a single system. The flexibility that ATM offers incorporates both circuit and packet switching techniques into one technology.

Althos Publishing, 404 Wake Chapel Road, Fuquay NC 27526 USA
1-919-557-2260 1-800-227-9681 Fax 1-919-557-2261 WWW.AlthosBooks.com

wireless Markup Language (WML)

ISBN: 0-9742787-5-0   **Price:** $34.99
Author: Bill Routt   **#Pages:** 292   **Copyright Year:** 2004

Wireless Markup Language (WML) Scripting, Scripting and Programming using WML, cHTML, and xHTML explains the necessary programming that allows web pages and other Internet information to display and be controlled by mobile telephones and PDAs.

Introduction to Bluetooth

ISBN: 0-9746943-5-5   **Price:** $14.99
Author: Lawrence Harte   **#Pages:** 60   **Copyright Year:** 2004

Introduction to Bluetooth explains what is Bluetooth technology and why it is important for so many types of consumer electronics devices. Since it was first officially standardized in 1999, the Bluetooth market has grown to more than 35 million devices per year. You will find out how Bluetooth devices can automatically locate nearby Bluetooth devices, authenticates them, discover their capabilities, and the process used to setup connections with them.

Introduction to CDMA

ISBN: 1-9328130-5-5   **Price:** $14.99
Author: Lawrence Harte   **#Pages:** 52   **Copyright Year:** 2004

Introduction to CDMA book explains the basic technical components and operation of CDMA IS-95 and CDMA2000 systems and technologies. You will learn the physical radio channel structures of the CDMA systems along with the basic frame and slot structures.

Introduction to Mobile Telephone

ISBN: 0-9746943-2-0   **Price:** $10.99
Author: Lawrence Harte   **#Pages:** 48   **Copyright Year:** 2004

Introduction to Mobile Telephone explains the different types of mobile telephone technologies and systems from 1st generation analog to 3rd generation digital broadband. It describes the basics of how they operate, the different types of wireless voice, data and information services, key commercial systems, and typical revenues/costs of these services. Mobile telephone technologies, systems, and services have dramatically changed over the past 2 years. tems to offer new services.

Introduction to Wireless Billing

ISBN: 0-9746943-8-X   **Price:** $14.99
Author: Avi Ofrane, Lawrence Harte   **#Pages:** 48   **Copyright Year:** 2004

Introduction to Wireless Billing explains billing system operation for wireless systems, how these billing systems are a bit different than traditional billing systems, and how these systems are changing to permit billing of non-traditional products and services. This book explains how companies bill for wireless voice, data, and information services.

Althos Publishing, 404 Wake Chapel Road, Fuquay NC 27526 USA
1-919-557-2260 1-800-227-9681 Fax 1-919-557-2261 WWW.AlthosBooks.com

ORDER FORM

Phone: 919-557-2260
800-227-9681
Fax: 919-557-2261   **Date:** _____
404 Wake Chapel Rd., Fuquay-Varina, NC 27526 USA
Email: success@Althos.com web: www.ALTHOS.com

Name: _____

Title: _____

Company: _____

Shipping Address: _____

City: _____ State: _____ Zip: _____

Billing Address: _____

City: _____ State: _____ Zip _____

Telephone: _____ Fax: _____

Email: _____

Purchase Order # _____ (New accounts, please call for approval)

Payment (select): VISA ____ AMEX ____ MC ____ Check ____

Credit Card #: _____ Expiration Date: _____

Exact Name on Card: _____

Qty.	BOOK #	ISBN #	TITLE	PRICE EA	TOTAL
Book Total:					
Discounts:					
Sales Tax (North Carolina Residents please add 7% sales tax)					
Shipping: Please apply accurate shipping rates and surcharge per client and sales.					
Total order:					

LaVergne, TN USA
17 August 2009
155050LV00002B/57/A